MODERN AIRCRAFT SERIES

Edited by Joe Christy

A new series of popular-priced books on aircraft and their operation for everyone interested in privately owned planes. Each volume is written by an expert in the field, and is printed on fine, white paper and profusely illustrated with photographs and diagrams. Each volume $2.95.

The Single-Engine Beechcrafts
Parachuting for Sport
Guide to Antique Planes
Cockpit Navigation Guide
Used Plane Buying Guide
Air Traffic Control
Modern Aerobatics
Classic Biplanes
Classic Military Biplanes
Aircraft Armament
Pilot's Weather Guide
Computer Guide
Club Flying
Guide to Homebuilts
Fighter Aircraft Pocketbook
All About Helicopters
Aces & Planes of WWI
Your FAA Flight Exam
Racing Planes Guide
Lightplane Engine Guide
The Piper Cub Story
Aircraft Dope & Fabric
Agricultural Aviation
Your Pilot's License
Instrument Flying Guide
Soaring Guide
Classic Monoplanes
Your Jet Pilot Rating
Lightplane Construction & Repair
Taking Pictures From The Air
Multiengine Flying
Airmanship After Solo
Legal Guide for Pilots & Owners
The Single-Engine Cessnas
Nav/Com Guide for Pilots

SPORTS CAR PRESS

Distributed by CROWN PUBLISHERS

419 Park Avenue South, New York, N.Y. 10016

THE SINGLE-ENGINE CESSNAS

JOE CHRISTY

New York
MODERN AIRCRAFT SERIES
A Division of Sports Car Press

Library of Congress Catalog Card Number: 71-147867

Published in New York by Sports Car Press, Ltd., and
simultaneously in Ontario, Canada, by
General Publishing Co.

ISBN 0-87112-037-2

Manufactured in the United States of America
by Semline Incorporated

CONTENTS

ACKNOWLEDGMENT

The author is grateful to Cessna's Jerry Kell, Director of Public Relations, and Larry Wiggins, Photo Editor, for their cheerful help during the preparation of this book. We're especially indebted to Jerry for compiling the Cessna Model Comparison Tables that follow Chapter Ten.

J.C.

FOREWORD

Clyde Vernon Cessna was a 31 year-old automobile mechanic in 1910 when he saw a trio of magnificent men perform in their flying machines at Oklahoma City. Clyde made detailed sketches of the Bleroit monoplane possessed by one of the birdmen and, a year later, had built a passable replica which he flew – after twelve attempts and eleven crashes – from the Salt Plains near Jet, Oklahoma. Clyde thereupon forsook the horseless carriage business and embarked upon an aerial exhibition tour of the county fair circuit. His first appearance was at Cherokee, Oklahoma, for which he received $300 for simply taking-off and circling the pasture. Such an accomplishment, in 1911, was unquestionably worth every cent of it.

During the next six years, until WW-I intervened, Cessna became a highly successful exhibition pilot (the *only* way there was to make a living with an airplane in those days), and he built an improved craft each winter at the family farm near Rago, Kansas. By 1917, his *Comet* monoplane averaged 124.62 mph in a flight between Wichita, Kansas and Blackwell, Oklahoma.

Clyde was 37 years old when America entered WW-I and was not accepted for military service. He farmed until 1919, then returned to the air. Late in 1924 he was still barnstorming (though by then using a Laird Swallow – sold to him by a Swallow salesman named Walter Herschel Beech) when Walter Beech and another Swallow employee, Lloyd Carlton Stearman, talked Cessna into joining them in the formation of a new airplane manufacturing business. With $10,000 in capital put up by Beech and Cessna, and Stearman's drawings of a three-place biplane, the Travel Air Company began operation with six employees in a small shop behind the Broadview Hotel in downtown Wichita.

But three chiefs in a single tepee made at least two too many. Though Travel Air grew rapidly, Stearman pulled out in 1926 to form his own company (later absorbed by Boeing), and Cessna

Dwane Wallace and his uncle, Clyde Cessna.
Photo taken shortly before Cessna's death in 1954.

sold out the following year to found the Cessna Aircraft Company (after several months as the Cessna-Roos Aircraft Corporation in partnership with Victor H. Roos, a former associate of Giuseppe M. Bellanca). Beech continued successfully with Travel Air, which was merged into the (then) giant Curtiss-Wright complex in 1929.

Meanwhile, Cessna steadily built his company. In 1929, he purchased 80 acres at Cessna's present main plant site and constructed five buildings. He had 80 employees, and his products were the Cessna Model A series; clean, four-place cantilever monoplanes powered with a variety of engines, the most popular of which was the Model AW, fitted with a 125-hp Warner or a 200-hp Wright Whirlwind. These Cessnas were Clyde's own designs, and were so basically sound one can still see some Model AW in single-engine Cessnas today. Between 1931 and 1936, Cessnas won the Detroit News "World's Most Efficient Airplane" trophy three times to gain permanent possession of that famed award.

The Great Depression halted production at Cessna in 1932, and one can only speculate as to what may have happened to the company (Clyde was nearing retirement age, and other airframe makers were being liquidated almost daily as the economic disaster worsened) had not Clyde given an airplane ride to his 12 year-old nephew, Dwane L. Wallace, back in 1923. From that time forward, young Wallace knew that his future would be in aviation. He soloed an OX-5 Travel Air in 1932, and graduated from Wichita State University in 1933 with a bachelor's degree in aeronautical engineering.

Since Cessna was closed down, Dwane worked briefly for Walter Beech (who had started Beech Aircraft Corporation in 1932, offering the Model 17 "Staggerwing"), then Dwane persuaded his uncle to reorganize the Cessna Aircraft Company and re-open the plant, producing a new four-to-five-place monoplane of outstanding performance. The resulting Cessna Airmaster kept the company alive and is regarded as a classic design today.

Clyde retired to his farm in 1936 (he died in 1954), and Dwane Wallace – at age 25 – became president of the company, as well as part-time janitor, test pilot, engineer and chief salesman. In fact, prize money won by Wallace racing the new C-34 Airmaster usually went to meet the company's modest payroll. Production averaged less than 30 units per year for 1934 through 1938.

The coming aircraft production demands of WW-II signaled an abrupt change in Cessna's fortunes. The company's first twin-engine plane, the T-50, appeared in 1939 and was bought first by the RCAF, then by the USAAF. More than 5,400 of these craft were built during WW-II as the AT-8, AT-17 and UC-78. The British and Canadians called it the "Crane;" the USAAF designation was "Bobcat." But to the cadets who transitioned in this twin-engine trainer it was affectionately (if irreverently) known as the "Useless-78," "Bamboo Bomber" or "Rhapsody in Glue."

Also during WW-II, Cessna produced troop-carrying gliders, and sub-assemblies for Boeing and Douglas – with, of course, greatly expanded factory facilities.

When the war ended, the company again turned to the commercial aircraft business. After a five-month conversion, Cessna introduced its first post-war models, the two-place 120 and 140. Production of these neat little craft reached 30 units per day in the summer of 1946, and that year's sales totaled nearly 4,000 airplanes.

Cessna broadened its market in 1947 with introduction of the four-place 190 and 195 – essentially, all-metal Airmasters. And the following year production was started on the new four-place Model 170.

C-165 Airmaster; from 1935's C-34 [illegible]
1941's C-165D 185 were constructed.

The L-19 Bird Dog (later designated O-1 by the Army) appeared in 1949, then the Korean War claimed a large portion of Cessna's production capacity with sub-assemblies for Boeing, Lockheed and Republic. But in 1954, along with the T-37 jet trainer, the five-place twin-engine Model 310 was introduced to the public. The tricycle-gear Models 172 and 182 came out in 1955, and then, when the two-place tri-gear Model 150 was added in 1959, the basic Cessna line of single-engine aircraft was established for years to come.

That was the year that Cessna expanded into the electronics business with acquisition of Aircraft Radio Corporation (ARC), and the year that total company sales for the first time passed the $100 million mark. In 1969 total sales were $282.9 million. And it is probably worth remarking that, in 1939 when Dwane Wallace closed the deal with Canada for the first 180 Cranes, there was exactly $5.03 in the company's bank account. As this is written, Dwane is still top man at Cessna as board chairman – and it's safe to say the company's bank account is somewhat larger these days.

CHAPTER 1: CESSNA 120 AND 140

During the fall of 1970, we checked with a number of aircraft dealers in Texas, Oklahoma and Kansas to get some impressions of the used plane market. We knew of course that new lightplane sales were down at the time, along with everything else in the aerospace industry, but we suspected that "experienced" lightplanes were still finding plenty of buyers.

This seemed to be the case. Every dealer we talked with said he could sell more used airplanes if he had them – and the two-place Cessna 140 and four-place Cessna 170 were the craft most often mentioned as desirable merchandise from both sellers' and buyers' points of view.

This meant that the fabric-wing 140, which sold for a maximum of $3,495 new (1946-1949), was bringing from $2,000 to $2,400 almost 25 years later. And a good metal-wing 140A (or a 140 with metalized wing), which had an original list price of $3,695, was commanding up to $2,750 in the 1970 market.

Simple inflation was largely responsible for the apparently negligible depreciation of low-priced lightplanes during these years. For example, shortly before we left the airport business in mid-1963, among the planes we sold were a clean little 140, for $2,100, and an equally good 170 for $4,700. Those same airplanes (in the same condition) would have brought at least that much seven years later.

But in addition to inflation, the basic soundness of the early post-war Cessnas has also contributed to their high re-sale value. The 120/140 is a trim little ship with honest flying characteristics and performance matching or exceeding that of the later 150.

The 120 and 140 are basically the same airplane; the 120 being the economy version without electrical system, starter or flaps, and possessing a minimum of instruments. Both aircraft are powered with the C-85 Continental engine – except for the 521 metal-winged 140A's produced at the factory. The 140A is fitted with a Continental C-90. Since many fabric-wing 140's were later converted to metal-wingers, these will probably still be equipped with the 85-hp engine and therefore are not 140A's. As originally delivered from

The Cessna 140 of 1946 had fabric-covered wing, 85-hp engine. Maximum list price was $3,495.

Cessna, all 120's and 140's were unpainted, except of course for silver dope finish on the fabric wings. In recent years, however, many have been given modern paint schemes.

Some 120's have been fitted with the Continental C-85-12 (the 140 engine, which has starter and generator) and a complete electrical system; and though such a craft may be called a "140," it really isn't – unless it has also been given the 140 wings with flaps and the 140's plusher cabin and extra instruments. But it does come close, because the 140's flaps are not the big "para-lift" barn-door airfoils found on later Cessnas, and do not help much, if any, on short take-offs, and though useful for landings, will seem anemic if you've been flying, say, a 150.

At the end of the sixties, there were about 1,200 120's and 3,500 140's still around.

Entering the 140 will at first seem an awkward and ungraceful exercise to those unaccustomed to the older tail-draggers, but it's no worse than a T-Craft, Luscombe or Aeronca, and certainly beats boarding a J-3 Cub. The seats are comfortable, but the cabin is small and becomes cramped for two (sitting side-by-side) if both are over 175 pounds.

Starter, throttle, parking brake, carburetor heat and mixture control are at bottom center of the instrument panel, with electrical switches immediately beneath. Elevator trim, flap control and fuel selector switch are between the seats. The fuel selector switch is three-position, for either of the two 12½-gallon wing tanks, or "Off." Fuel gauges are of the direct-reading type and located in each wing root inside the cabin. Fuel to the engine is gravity-fed via the selector-switch junction, then through a strainer on the firewall to the carburetor.

These little Continental engines are easy to start (when properly maintained), though on a cold day three or four strokes of the primer are required. Since this is a tail-wheel airplane, taxi in an S-pattern to clear the blind spot over the nose. During the pre-take-off check, test mags at 1,800 rpms; set carb heat at "Cold," turn elevator trim wheel to "Take-Off;" determine that flaps are up and locked, and of course be sure your fuel selector is turned to the fullest tank. This is also when you are supposed to lock the door and windows, although we can't guarantee that they'll stay that way on some 140's after all these years.

Average take-off roll with the 140 at full gross load (1,450 lbs) is between 600 and 750 feet in a no-wind condition, depending upon ground elevation and temperature. Holding the tail a little below horizontal, she breaks ground at slightly over 40 mph. Then, climb-out should be made at between 80 and 90 mph for good engine cooling. This will give you a rate-of-climb of about 600 feet per minute initially; and on a normal day, somewhere between 300 and 400 fpm at 7,000 ft.

Sure, this is less than fighter airplane performance; but you are aviating with 85 horsepower and burning only 5 gallons of gasoline per hour. At the 140's normal cruising speed of 105 mph, that's more than 20 miles per gallon – which is far better than your car (any car) can do. Full Throttle (2,575 rpms) will up your speed to between 115 and 120 mph and raise gas consumption to about 6 gallons per hour.

Stalls are gentle in the 120 and 140. We once read a "pilot's report" in an aviation magazine wherein the writer said the 140 "stalls without warning." He just wasn't payin' attention. This craft does have a marked buffett preceding the stall. The nose falls through at about 47 mph, power off and clean. With flaps and power off, the break comes at about 43 or 44 mph. Flaps down and power on, she quits flying at about 38 mph. Recovery is quick, and there is some aileron control into the stall, although the 140's ailerons – like those on other single-engine Cessnas – are a bit too soft to suit some pilots.

Landing the 140, use carburetor heat and set the mixture control at "Full Rich." Sixty-five to seventy mph is a good gliding speed, with some nose-up trim cranked into lighten the control pressure. Flaps will steepen your glide without a build-up in airspeed, and do help to sink through the ground cushion after flareout. The 120, which has no flaps, seems to want to float a little at this point. But both 120 and 140 are fairly easy to land, and each has a relatively wide tread for planes of this configuration (the first of the Cessna steel-spring landing gears), and therefore are surefooted on the ground.

In sum, it's hard to imagine how anyone could get into the air with his own airplane for much less money than with a 120 or 140 – assuming that you find one that has been well maintained and isn't on the brink of expensive refurbishing. The only FAA Bulletin

The 1949 Cessna 140A had metal wing, single strut, and 90-hp engine; sold for $3,695. Cruising speed, 105 mph. Still popular on the used plane market.

CESSNA 140

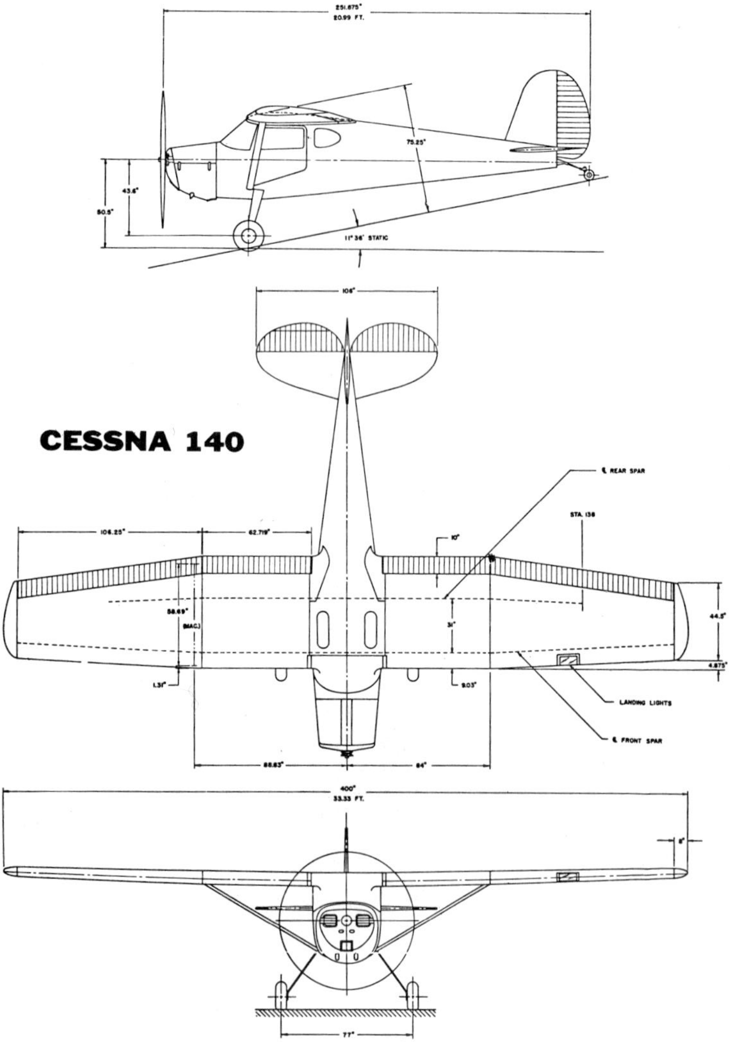

251.875"
20.99 FT.
75.25"
43.6"
50.5"
11° 36' STATIC
106"
106.25"
62.719"
℄ REAR SPAR
STA. 138
10"
31"
44.5"
4.875"
1.31"
9.03"
LANDING LIGHTS
℄ FRONT SPAR
88.83"
84"
400"
33.33 FT.
8"
77"

(AD Note) of consequence ever issued on these planes was AD-50-31-1, which calls for a reinforcement to the vertical fin spar. So, if you're thinking of buying such a plane, check the airframe log to make sure this has been complied with.

You'll also want to make a close inspection for corrosion and determine the condition of the wing fabric among other things. But let's face it, there is really only one reasonably safe way to select a used airplane of any kind: hire an experienced A&P (airframe & powerplant) mechanic – one who has no interest in the deal one way on another – to thoroughly inspect a plane before you buy it. This will cost you about $50.

Yes, it hurts a little to ante up $50; particularly if you go home without that airplane you had your heart set on. But it's better than buying a machine that spends much of its time in the shop running up repair bills.

SPECIFICATIONS, CESSNA 120 & 140

Engine	Continental C-85 or C-90
Wingspan	33 ft 4 in
Length	21 ft 6 in
Height	6 ft 3 in
Wing Area	159.3 sq ft
Wing Loading	9.1 lbs sq ft
Power Loading	17.1 lbs per hp
Empty Weight	120, 770; 140, 890 lb
Gross Weight	1,450 lbs
Useful Load	120, 680; 140, 560 lb
Fuel Capacity	25 gal
Oil Capacity	5 qts

PERFORMANCE, CESSNA 120 & 140*

Take-Off Distance	500 ft
Rate-of-Climb (Initial)	680 ft per min
Service Ceiling	15,500 ft
Top Speed	125 mph
Cruising Speed	105 mph
Range (with reserve)	450 mi
Stall Speed (without power; flaps up)	49 mph
Stall Speed (without power; flaps down)	45 mph
Landing roll (without brakes)	230 ft

*Original factory figures computed at full gross load and standard atmosphere (sea level; 59 degrees F).

CHAPTER 2: CESSNA 190 AND 195

Our description of the Cessna 190 and 195 will undoubtedly be a little biased, but it's not our fault. We don't *start* feuds with airplanes or other inanimate (?) objects. But "Shakey Jake" clearly disliked us from the beginning. Shakey was a 195 who belonged to a flying club. He got his name from his Jacobs engine, and he was brought to us for sundry maintenance and repairs. That is, he was brought to our shops when we were in the FBO business several years ago.*

Shakey was in and out of our place for many months, during which time he humiliated this writer on many occasions, dripped oil on our clean floors and obviously would have done us bodily harm had he ever got the chance. He had a crosswind landing gear, and everytime he saw us approach him he'd splay his wheels in different directions and dare us to move him. Everyone else around there could roll Shakey to or from the hangars and he'd go quietly. But not for this reporter; he hated us.

Shakey got the best of us in the end. His owners brought him in for a new windshield and, innocently, we gave an estimate and agreed to do the job. Well, please be advised, friends, such a request should be regarded in law as sufficient provocation for justifiable homicide. At least, when it involves an airplane with Shakey's delinquent social tendencies.

As it happened, our AI was away on a ferry flight, our number one A&P was on vacation, and there was no one in the shops except a young A&P just out of vo-tech school. And it quickly became apparent that replacing a 195 windshield was a job for two skilled airframe mechanics. But, naturally, Shakey would not have needed a windshield had our experienced men been on hand. Also, as you would expect, this happened in the middle of an Oklahoma summer. So, we sat in Shakey's cabin—where the temperature approximated that of Death Valley—for a couple of days, and tried to guess which of the roughly 4,000 locknuts we were supposed to hold while the young A&P tightened from the outside and the suspicion grew that our replacement windshield was really for a DC-3 or something else. Anyway, our AI returned at the end of the

*The author is not an aircraft mechanic. As part-owner of this flight operation, his principal duties, it seems, were often those of line boy, mechanics' helper and rest room cleaner.

week, looked at the job in amazement, and did it all over again. We probably only lost a few hundred dollars on the deal.

With this as background, you'll understand if we are less than enthusiastic about the 195. As we said at the start, it clearly isn't our fault. Any pilot will agree that some airplanes just naturally dislike some people.

The first 190 appeared in 1947 and was powered with the Continental R-670-23 engine of 240 hp. The 195 quickly followed, and was offered with a choice of three Jacobs powerplants: the 245-hp R-755-9; the 275-hp R-755-B2 and the 300-hp R-755-A2.

A total of 204 190's were built, 1947-1953 inclusive; and 890 195's were produced, 1947-1957 inclusive. In addition, 83 of the 300-hp 195's were built for the military as the LC-126, 1950-1952, with 15 going to the Air Force, 5 to the Air National Guard and 63 to the Army (our nemesis, Shakey, was actually an Army surplus LC-126).

The Cessna 195

Prices ranged from $12,750 for early models up to $24,700 for the last 300-hp 195's. About 450 Cessna 195's were still on the FAA's "active" list in 1970; and some could be found on the used airplane market for $6,000 to $7,000 in "average" condition. It should be noted, however, that this craft possesses an impressive number of staunch friends, enough to form an International 195 Club, and these owners, flying mint-conidtion 195's, do not of course place "average" prices on them – if and when they are willing to sell at all. Thus, there were some like-new (or better-than-new) 195's around valued from $10,000 upwards at the beginning of the seventies.

Most 195's have a retracting step that extends when the cabin door is opened; on some this step may be permanently fixed. The cabin is roomy — positively spacious by lightplane standards — with a pair of individual adjustable seats in front and a solid-cushion type seat in back that accommodates three. As is obvious from the outside, forward visibility over the nose, from the pilots' seats, is very limited when this airplane is in three-point attitude. While on the ground, an acre or two of instrument panel seems to block out most of the world, so taxiing must be done in an S-pattern if you want a clue as to what you're about to run in to.

Fuel from the 40-gallon tank in each wing (total: 80 gal.) is gravity-fed to the Continental engine in the 190; but a fuel pump is used in the 195 because of its higher horsepower. In case of fuel-pump failure, a by-pass line will keep the Jacobs supplied with sufficient fuel by gravity alone to maintain normal cruise. The Jacobs must be primed on every start; two or three strokes of the primer on warm days, and five to seven strokes on cold days.

Starting the Jacobs, fuel mixture of course is set at "Full Rich;" carburetor heat at "Cold," but the propeller is set at "Low RPM" to prevent taking too much oil from the engine at this point. With the master switch "On," the engine primed and throttle cracked, turn the key ignition to "Battery Start" position, make sure your prop is clear and then press the starter button. As soon as the engine fires, turn the key ignition to "Both Run." The engine may require another shot or two of prime to keep it running at first.

By this time, the roomy transport-type cabin, generous expanse of instrument panel and that big radial barking defiance at the world, may combine to give you the air of Captain Sternjaw, departing upon his regular run to Kalamazoo. If so, okay, because there are times when this airplane needs full attention and a bit of

More than 120 members of the International 195 Club met at Wichita in 1970, flying in a total of 41 Cessna 195's for the three-day party.

professionalism in flying technique, especially after landing flare-out in a cross-wind, when it has a strong tendency to weathercock.

During the take-off run, visibility over the nose improves as the tail comes up. Full right rudder trim will compensate for much of the torque, and the 195 will fly off at about 60 mph – with an average take-off roll of about 900 feet – in a slightly tail-low attitude.

Full throttle is allowable for one minute, and this will give you an initial climb-rate of about 1,000 fpm at 2,200 rpms and 85 mph. But the recommended climb/cruise configuration is 2,000 rpms at 23 inches of manifold pressure with about 100 knots or 115 mph showing. This produces, usually, around 500 to 600 fpm, and the 195 will hold this, with very little drop in the rate-of-climb, through 4,000 feet.

Althuogh a lot of people have described the 195 as a "good instrument airplane," we'd have to disagree. True, good turns (without altitude loss or gain) are easier on instruments until one learns where the horizon should be with relation to the top of the instrument panel, or whatever you finally pick as a reference for eyeball turns; and since aileron response is slow there's naturally less over-controlling with a set of "white knuckles" on the wheel during instrument flight. The 195 does have a big airplane feel, which alone must induce a certain feeling of security. But it needs to be flown. In cruise configuration, no matter how carefully you trim it, the plane will, left to itself, slowly porpoise, alternately gaining and losing up to 75 feet or even more.

Stall behavior is neither good nor bad. The 195 has plenty of aileron into the stall, and you'll use it. Power off and clean, the stall comes at about 65 mph. With power off and flaps, the break is a bit sharper at about 62 mph. With power on and flaps down, the pay-off registers at about 56-58 mph. These figures are average, for a moderate day and at near full gross load.

This airplane is easy to land – if the wind is straight down the runway. In a cross-wind some artful rudder work is required and, as far as we were ever able to discover, a certain amount of good luck.

This trait undoubtedly prompted installation of a cross-wind landing gear on some 195's (and 170's). And we'll say this for the cross-wind gear: it does exactly what it's supposed to do. That is, it keeps you rolling straight down the runway while the airplane

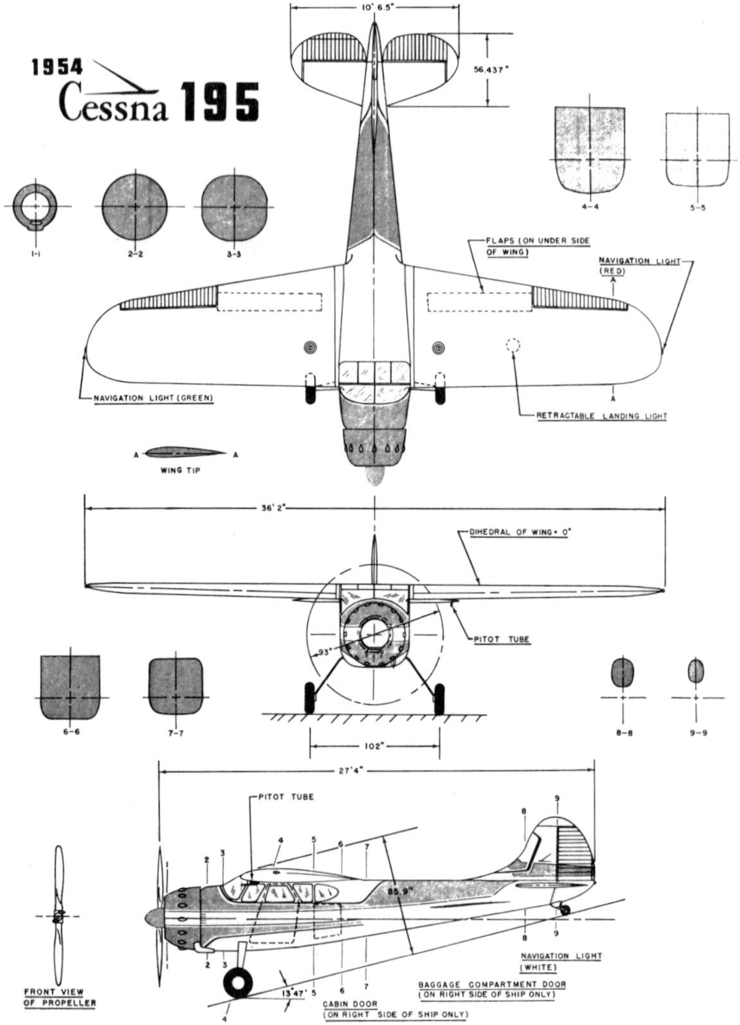
1954
Cessna 195
10' 6.5"
56.437"
1-1
2-2
3-3
4-4
5-5
FLAPS (ON UNDER SIDE OF WING)
NAVIGATION LIGHT (RED)
NAVIGATION LIGHT (GREEN)
RETRACTABLE LANDING LIGHT
A
A
WING TIP
36' 2"
DIHEDRAL OF WING • 0°
PITOT TUBE
93"
6-6
7-7
8-8
9-9
102"
27'4"
PITOT TUBE
85.9"
NAVIGATION LIGHT (WHITE)
BAGGAGE COMPARTMENT DOOR (ON RIGHT SIDE OF SHIP ONLY)
13°47'
CABIN DOOR (ON RIGHT SIDE OF SHIP ONLY)
FRONT VIEW OF PROPELLER

points in another direction. But it does take a modicum of faith to use it at first. Every nerve in your being will rebel at the prospect of actually touching-down while still crabbing 15 degrees or so into the wind. After a few landings, however, you'll probably feel it's the only practical gear for the 195. That is, until you have to wrestle the stubborn thing in and out of the hangar a few times.

Depending upon its condition, the 195 should give you a true air speed of 165-170 mph at 7,000 feet where about 70% of its power is available. Fuel consumption is about 16 gallons per hour with the 300-hp Jacobs.

Like we said in the beginning, this isn't a very objective report – maybe even a little unfair. But if an airplane ever gets it in for you, the way Shakey did for us, you'll understand.

SPECIFICATIONS, CESSNA 195

Engine	Jacobs R-755
Propeller	Hamilton Standard Constant Speed
Wing Span	36 ft 2 in
Length	27 ft 4 in
Height	7 ft 2 in
Wing Loading	15.36 lbs per sq ft
Power Loading	11.16 lbs per hp
Empty Weight	2,100 lbs
Gross Weight	3,350 lbs
Useful Load	1,250 lbs
Baggage Capacity	220 lbs
Fuel Capacity (usable)	75 gal
Oil Capacity	5 gal

PERFORMANCE, CESSNA 195

Take-Off Distance	800 ft
Take-Off over 50-ft Obstacle	1,600 ft
Landing Distance	700 ft
Landing over 50-ft Obstacle	1,600 ft
Rate of Climb, Maximum Initial	1,200 ft per min
Service Ceiling	18,300 ft
Maximum Speed	185 mph
Cruise Speed (70% Power)	170 mph
Range at 70% Power	800 mi
Stall Speed, Power Off, Flaps 45 Degrees	62 mph

The first (1948) Cessna 170 had two wing struts on each side, fabric-covered wing and a scaled-up 140 tail. Price was $7,245.

CHAPTER 3: CESSNA 170

The Cessna 170 series airplanes are still much in demand on the used market more than 20 years after this design first appeared. Good ones, late in 1970, were bringing 75% of their original purchase price. New prices ranged from $7,245 to $8,295 during production (1948-1955 inclusive), and a 170A or 170B in good condition would quickly find a buyer willing to pay up to $5,000 for it in 1970.

As with the used 140's, a portion of this relatively-high figure could be attributed to inflation. But again, as with the 140's, the Cessna 170 series compare very favorably, performance-wise, with the posher and sleeker tri-cycle gear models that replaced them. A comparison with the 1971 Model 172 (basic price, $13,425), will show what we mean:

	170B (1952)	172 (1971)
Top Speed	140 mph	139 mph
Cruising Speed	120 mph (65% power)	131 mph (75% power)
Initial Climb Rate	690 fpm	645 fpm
Useful Load	995 lbs	1,050 lbs
Range	540 mi	615 mi

To be fair, we must point out that the above figures do not reflect the 172's added comfort, quieter cabin, easier landing and ground handling, 360-degree vision and a number of other little niceties, including a rakishly-swept rudder (which adds nothing except looks, and probably even costs a couple of mph in speed). And considering the additional fact that the 172 is brand new, while the 170B probably has at least 4,000 hours on its aging frame, there's little doubt but that the new 172 is a better buy at perhaps three times the price (including avionics) – if one has the money and/or a sound financial reason for operating a plane in this class.

But for the pilot who wants a four-placer primarily for pleasure, and cannot or does not wish to invest more than \$4,000-\$5,000 in such a machine, a well-maintained 170 may well be his best choice. There is also the possibility of a used 172, and we'll discuss that series in a following chapter.

The original 170 had a fabric-covered wing, two wing struts on each side and a 140-type rudder. The 170A, which appeared in 1949, had a metal-covered wing and new dorsal fin, plus a single lift strut on each side. The 170B, similar to the 170A but with L-19 type flaps added, was introduced in 1952. Officially, production ceased with the 1955 Model 170B; but 72 were built in 1956 (alongside the new 172), and 36 were produced in 1957, according to Cessna records. Altogether, 5,136 of the 170 series were built, and about 2,800 remained on the FAA's active list at the end of 1970.

The 170 is fairly easy to board for a tail-wheel airplane, and visibility over the nose is good. Noise-level in flight is about what you'd expect in an older airplane. This craft does want to weathercock on the ground in a good wind, and that can be a problem on a big airport when you have a long way to taxi. Having said that, it's hard to find any other fault with the 170. Its flying characteristics are all good; it's a pleasure to fly.

Cessna 170B of 1952 was all metal and had performance matching or surpassing the Model 172 which descended from it. Priced at \$8,295 when production stopped in 1955, this model, in good condition, would bring 3/4ths of that in the used market in 1971.

The venerable L-19 Bird Dog (0-1), kissin' cousin to the 170, has maximum speed of 116 mph at 5,000 ft. Cessna produced 3,431 Bird Dogs from 1950 through 1969 inclusive.

Following standard pre-take-off procedures, the 170B will normally become airborne in about 700-800 feet, depending upon all-up weight and air density, indicating around 55 mph. Initial climb will be in the neighborhood of 700 fpm; up to 1,000 fpm if you are light.

This is a stable airplane in all axes, and requires reasonable coordination between wheel and rudder pedals for good turns. The 170's 145-hp Continental engine delivers 65% power at 5,000 feet pressure altitude turning 2,450 rpms, and produces 120 mph TAS with 8 gph fuel consumption at this setting. Since later model Cessnas using a quite similar engine cruise it at 2,700 rpms, the 170's recommended cruise rpms should result in fewer mechanical problems.

Stalls are gentle in this craft, and there is no wing drop except

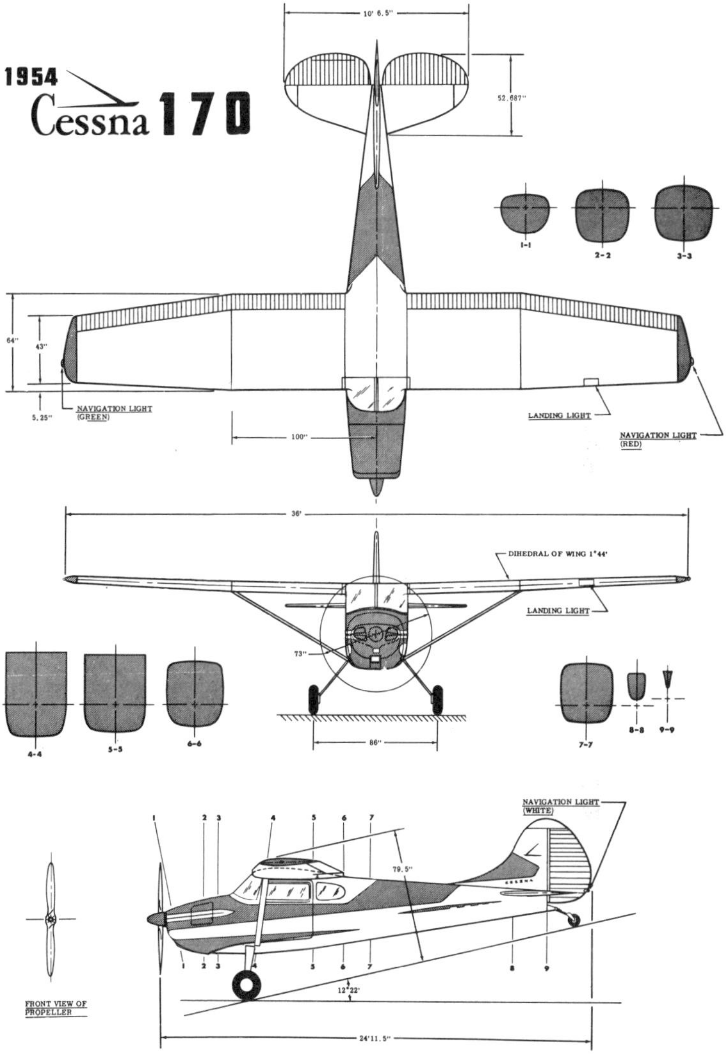
1954 Cessna 170
10' 6.5"
52.687"
1-1
2-2
3-3
64"
43"
5.25"
NAVIGATION LIGHT (GREEN)
100"
LANDING LIGHT
NAVIGATION LIGHT (RED)
36'
DIHEDRAL OF WING 1°44'
LANDING LIGHT
73"
86"
4-4
5-5
6-6
7-7
8-8
9-9
NAVIGATION LIGHT (WHITE)
79.5"
12°22'
24'11.5"
FRONT VIEW OF PROPELLER

in full-power stalls. In all cases, this machine recovers from the stall very quickly. We also regard the 170 as easy to land for a tail-dragger. It will float if you're carrying only a little too much speed; it was designed for full-stall landings. Still, the old pros tell us that's the way all lightplanes should be landed, nose wheel or not.

Summing up: The 170 is a simple and straightforward machine that is economical to operate and maintain. Its systems remind us of something we once heard Bill Lear say: "You'll never have to replace, repair or maintain anything you leave out." That was Lear's way of re-stating the engineers' hallowed axiom of "KISS" ("Keep It Simple, Stupid"), and a principle that's especially sensible applied to the design of light airplanes.

SPECIFICATIONS, CESSNA 170

Engine	Continental C-145
Propeller	McCauley IA170 or Sensenich 73BR-50
Wing Span	36 ft
Length	24 ft 11½ in
Wing Area	175 sq ft
Wing Loading	12.6 lbs per sq ft
Power Loading	15.2 lbs per hp
Height	6 ft 7½ in
Empty Weight	1,205 lbs
Gross Weight	2,200 lbs
Useful Load	995 lbs
Fuel Capacity	42 gal
Baggage Capacity	120 lbs

PERFORMANCE, CESSNA 170

Take-Off Distance	700 ft
Take-Off Distance Over 50 ft Obstacle	1,500 ft
Landing Distance	500 ft
Landing Distance Over 50 ft Obstacle	1,100 ft
Rate of Climb, Maximum Initial	690 ft per min
Service Ceiling	15,500 ft
Maximum Speed	140 mph
Cruise Speed (65% Power)	120 mph
Range at 65% Power	540 mi
Stall, Power Off, No Flaps	58 mph
Stall, Power Off, Flaps Down	52 mph

CHAPTER 4: CESSNA 180 AND 185

When the Cessna 180 was first introduced in 1953, it was enthusiastically received by businessmen pilots of that time, and by operators who needed a flying machine in this class with more flexibility than the 170: a nimble load-carrier with a cruise above 150 mph capable of flying from all kinds of fields, including those at high altitudes. Also a good instrument airplane, it was, in short, the flying machine that put the "U" in utility.

The early 180's grossed only 15% (350 lbs) more than the 170, but possessed almost 40% more horsepower (225-hp vs. 145-hp), and this resulted in a quick-footed craft.

The 180 lost its businessmen buyers, however, when Cessna decided, in 1956, to bring out the 182 (essentially, a 180 with an idiot landing gear). Nevertheless, the 180 continued in production because a lot of operators in the back-country still found it ideally suited to their requirements and, at the beginning of the seventies, it probably was the most widely used ranch and bush airplane in the world. This is why those of us whose activities take us from one paved runway to another in the "Lower 48," don't see many 180's; they're mostly working in Latin America, Alaska and the American West. More than 5,000 had been built at the end of the sixties.

The 180 hasn't changed an awful lot over the years. It received five extra horsepower (to 230-hp) in 1956 and its gross weight had grown to 2,800 lbs by 1970, along with increased range (bigger fuel tanks), and an extra window on each side. The 180's nose cowling was re-designed in 1959 for better cooling and additional ram-air pressure to the intake, and the gear legs have been re-shaped to obtain a wider track. In 1970, the plane appeared with new conical-cambered wingtips and a few fresh pretties in the cabin. By that time the price – pushed by inflation – was up to $19,725. In 1953, the first 180 had a list price of $12,950.

Since the 180 is a working airplane, its cabin is austere; no fancy upholstery, and its basic interior configuration includes only the pilot's seat. Removable seats for five more passengers are extra-cost items. Flaps are manually operated, and the stabilizer trim wheel and fuel selector switch are located between the seats. It is, in short, an airplane of few surprises – except in the performance

1968 Cessna Model 180

department. If you've been flying its citified counterpart (the 182), you'll find that the 180 out-performs the tri-gear dude in all areas except top speed.

Tail-wheel airplanes *do* require extra skill and experience on the ground and in cross-wind landings, and we fail to understand why some av-writers have claimed otherwise. But if you learned

1969 Cessna Skywagon 180

to fly in a tail-dragger, or have some experience in one, you are bound to like the 180. At anything less than full gross load, it'll climb out at 1,000 fpm or better, and its ceiling of almost 20,000 feet promises that it'll take you anywhere and be content with the merest excuse for a landing patch.

The 180 has excellent slow-flight characteristics with a solid control-feel, particularly rudder, and refuses to fully stall with power off and only one or two aboard. It merely shakes a little and bobs its nose seeking a flyable angle-of-attack while determinedly resisting a clean break. She will of course break from power stalls, or even if you are carrying enough weight, power or no power. Even then, this airplane is almost immediately flying again as soon as you relax back-pressure on the wheel.

Summing up: a delightful craft if you are one who prefers to fly the airplane rather than have the airplane fly you.

The Cessna 185 Skywagon is, essentially, a beefed-up 180 with more power. Externally, the airframe may appear to be the same as the 180's, but internally the 185 has additional bracing and a redesigned wing for extra strength. It also has a big dorsal fin which quickly distinguishes it from the 180. The 185 was added to the Cessna utility fleet in 1961. It was fitted with the 260-hp Continental I0-470-F engine through the 185E model of 1966. In that year, Cessna offered the E Model with the I0-520D 300-hp Continental, and in 1967 dropped the 260-hp 180 altogether. From 1966 through 1970, Cessna called the 300-hp model the A185E Skywagon.

Skywagon it is. With room for six big people, a useful load of 1,775 lbs (1,620-1,730 lbs for earlier models), and a cruising speed of 169 mph, the 185 sort of picks up where the 180 leaves off – like going from a half-ton to a three-quarter ton pickup, or perhaps more to the point, from a Chevy station wagon to a Buick wagon. It had good performance in 1961, and ten years later (unlike most airplanes) it had not taken on a lot of weight which would subtract from that performance. When the 40 extra horsepower was added, gross weight went up only 150 lbs to 3,350; and empty weight was increased but 15 lbs. Therefore, that extra power was available to do some useful work, and not wasted on a heavier airframe. To put it another way, the 185, which weighs only 30 lbs more than the 180 (empty), carries 520 lbs more load at about the same speed, 169 vs. 170 mph.

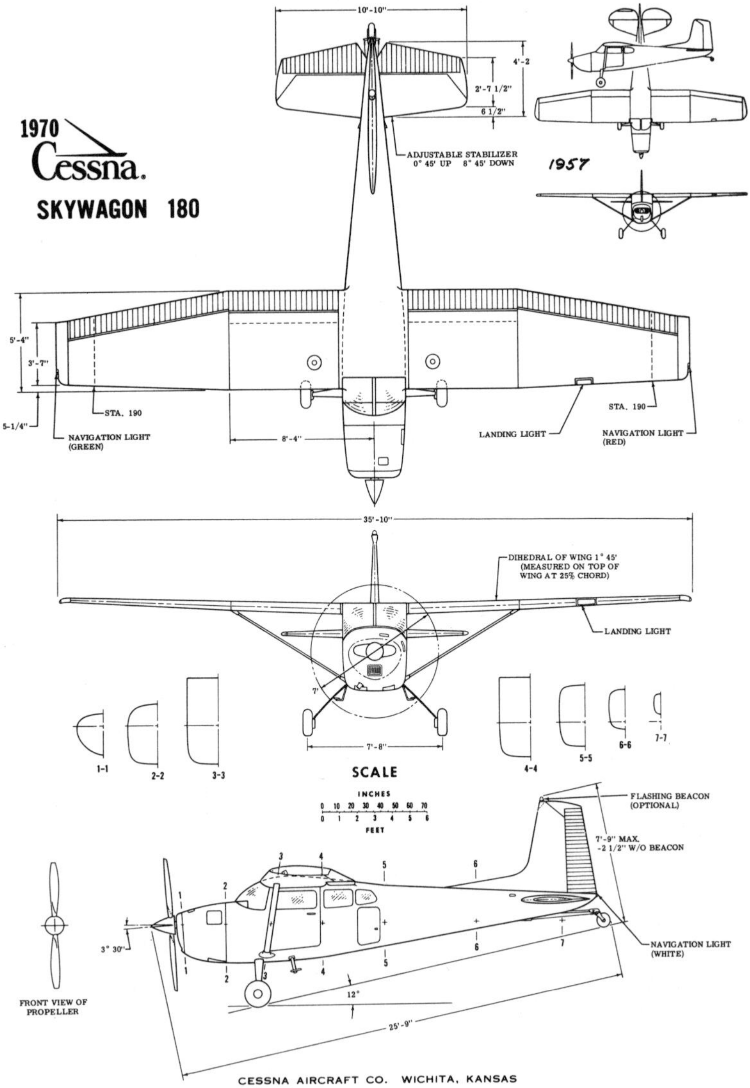
1970
Cessna.
SKYWAGON 180
10'-10"
4'-2
2'-7 1/2"
6 1/2"
ADJUSTABLE STABILIZER
0° 45' UP 8° 45' DOWN
1957
5'-4"
3'-7"
5-1/4"
STA. 190
NAVIGATION LIGHT
(GREEN)
8'-4"
STA. 190
LANDING LIGHT
NAVIGATION LIGHT
(RED)
35'-10"
DIHEDRAL OF WING 1° 45'
(MEASURED ON TOP OF
WING AT 25% CHORD)
LANDING LIGHT
7'
7'-8"
1-1
2-2
3-3
4-4
5-5
6-6
7-7
SCALE
INCHES
0 10 20 30 40 50 60 70
0 1 2 3 4 5 6
FEET
FLASHING BEACON
(OPTIONAL)
7'-9" MAX.
-2 1/2" W/O BEACON
NAVIGATION LIGHT
(WHITE)
3° 30'
12°
25'-9"
FRONT VIEW OF
PROPELLER
CESSNA AIRCRAFT CO. WICHITA, KANSAS

Guide for Estimating Cost of Operation—Skywagon 180

	*PER YEAR	PER YEAR	*PER HOUR	PER HOUR	PER HOUR	PER HOUR
FLIGHT HOURS PER YEAR.				300	500	700
FLIGHT MILES PER YEAR: 162 mph, 75% power at 6,500 ft., Lean Mixture.				48,600	81,000	113,400
GASOLINE: 14.0 gal./hr. at 43¢/gal.				$ 6.02	$ 6.02	$ 6.02
OIL: 1 pt./hr. plus oil change every 25 hrs. (12 qts. per change) at 63¢/qt.				$ .63	$ .63	$ .63
AIRPLANE & ENGINE MAINTENANCE.				$ 2.00	$ 2.00	$ 2.00
TOTAL OPERATIONAL COSTS (listed above).				$ 8.65	$ 8.65	$ 8.65
RESERVE FOR ENGINE OVERHAUL.				$ 1.54	$ 1.54	$ 1.54
STORAGE: at $50.00 per month.		$ 600		$ 2.00	$ 1.20	$ .86
INSURANCE (a guide for the following coverage): All risks hull coverage (Deductibles $250.00 in motion) (Not in motion $50.00).		$1,189				
Liability Coverage (Single limit $250,000) (Including passengers).		$ 140				
TOTAL INSURANCE (Industrial private business & pleasure use by a qualified pilot)		$1,329		$ 4.43	$ 2.66	$ 1.90
COST PER HOUR.				$16.62	$14.05	$12.95
COST PER MILE: 75% power, 162 mph.				$ .103	$ .087	$ .080
COST PER SEAT MILE: 6 People.				$ .017	$ .015	$ .013

FOR DEPRECIATION: Consult your tax authority. To estimate the per hour or per mile cost, divide annual depreciation amount by annual hours or miles.

*YOUR COSTS: No doubt vary from these national verages. This column is included for you to compute your estimate of your hours, miles and costs.

1970 Cessna Skywagon 180

1970 Skywagon 180 Performance and Specifications

	LANDPLANE	AMPHIBIAN	*FLOATPLANE	SKIPLANE
GROSS WEIGHT	2800 lbs. 1270 kg	2950 lbs. 1338 kg	2950 lbs. 1338 kg	2800 lbs. 1270 kg
SPEED:				
Top Speed at Sea Level	170 mph 274 kph	149 mph 240 kph	149 mph 240 kph	149 mph 240 kph
Cruise, 75% Power at 6500′	162 mph 261 kph	147 mph 237 kph	147 mph 237 kph	144 mph 232 kph
RANGE:				
Cruise, 75% Power at 6500′ 60 Gallons usable fuel No Reserve	695 mi. 1118 km 4.3 hrs. 162 mph 261 kph	630 mi. 1014 km 4.3 hrs. 147 mph 237 kph	630 mi. 1014 km 4.3 hrs. 147 mph 237 kph	620 mi. 998 km 4.3 hrs. 144 mph 232 kph
Cruise, 75% Power at 6500′ 79 Gallons usable fual No Reserve	925 mi. 1488 km 5.7 hrs. 162 mph 261 kph	835 mi. 1344 km 5.7 hrs. 147 mph 237 kph	835 mi. 1344 km 5.7 hrs. 147 mph 237 kph	820 mi. 1320 km 5.7 hrs. 144 mph 232 kph
Optimum Range at 10,000′ 79 Gallons usable fuel No Reserve	1215 mi. 1955 km 10.0 hrs. 121 mph 195 kph	1090 mi. 1754 km 9.5 hrs. 114 mph 183 kph	1090 mi. 1754 km 9.5 hrs. 114 mph 183 kph	955 mi. 1537 km 9.5 hrs. 101 mph 162 kph
RATE OF CLIMB AT SEA LEVEL	1090 fpm 332 mpm	990 fpm 302 mpm	990 fpm 302 mpm	
SERVICE CEILING	19,600 ft. 5974 m	16,000 ft. 4877 m	16,000 ft. 4877 m	
TAKEOFF:		On Land / On Water		
Ground Run (or Water Run)	625 ft. 191 m	On Land: 1360 ft. 415 m On Water: 1280 ft. 390 m	1280 ft. 390 m	
Total Distance Over 50′ Obstacle	1205 ft. 367 m	On Land: 2185 ft. 666 m On Water: 2070 ft. 631 m	2070 ft. 631 m	
LANDING:				
Ground Roll (or Water Run)	480 ft. 146 m	On Land: 1025 ft. 312 m On Water: 735 ft. 224 m	735 ft. 224 m	
Total Distance Over 50′ Obstacle	1365 ft. 416 m	On Land: 1490 ft. 454 m On Water: 1720 ft. 524 m	1720 ft. 524 m	
STALL SPEED:				
Flaps Up, Power Off	65 mph 105 kph	65 mph 105 kph	65 mph 105 kph	65 mph 105 kph
Flaps Down, Power Off	58 mph 93 kph	58 mph 93 kph	58 mph 93 kph	58 mph 93 kph
EMPTY WEIGHT: (Approximate) (One Seat)	1545 lbs. 701 kg	2100 lbs. 954 kg	1055 lbs. 841 kg	1690 lbs. 767 kg
USEFUL LOAD	1255 lbs. 569 kg	850 lbs. 384 kg	1095 lbs. 497 kg	1110 lbs. 503 kg
BAGGAGE: (Total Capacity)	400 lbs. 181 kg	400 lbs. 181 kg	400 lbs. 181 kg	400 lbs. 181 kg
WING LOADING	16.1 lbs/ft^2 78.6 kg/m^2	17.0 lbs/ft^2 82.6 kg/m^2	17.0 lbs/ft^2 82.6 kg/m^2	16.1 lbs/ft^2 78.6 kg/m^2
POWER LOADING	12.2 lbs/hp 5.5 kg/hp	12.8 lbs/hp 5.8 kg/hp	12.8 lbs/hp 5.8 kg/hp	12.2 lbs/hp 5.5 kg/hp
FUEL CAPACITY: Total				
Standard	65 gal. 246 liters	65 gal. 246 liters	65 gal. 246 liters	65 gal. 246 liters
Optional Long Range Tanks	84 gal. 318 liters	84 gal. 318 liters	84 gal. 318 liters	84 gal. 318 liters
OIL CAPACITY: Total	12 qts. 11.4 liters	12 qts. 11.4 liters	12 qts. 11.4 liters	12 qts. 11.4 liters
POWER	Six-Cylinder Engine, 230 Rated HP at 2600 RPM			
PROPELLER: Constant-Speed, Diameter	82 in. 2.08 m	88 in. 2.24 m	88 in. 2.24 m	82 in. 2.08 m
WING SPAN	35 ft 10 in 10.92 m	35 ft 10 in 10.92 m	35 ft 10 in 10.92 m	35 ft 10 in 10.92 m
WING AREA	174 $ft.^2$ 16.2 m^2	174 $ft.^2$ 16.2 m^2	174 $ft.^2$ 16.2 m^2	174 $ft.^2$ 16.2 m^2
LENGTH	25 ft 9 in 7.85 m	27 ft 6 in 8.39 m	27 ft. 8.24 m	25 ft 9 in 7.85 m
HEIGHT	7 ft 9 in 2.36 m	12 ft 8 in 3.86 m	12 ft 2 in 3.71 m	7 ft 9 in 2.36 m

*Edo Model 628-296 Floats

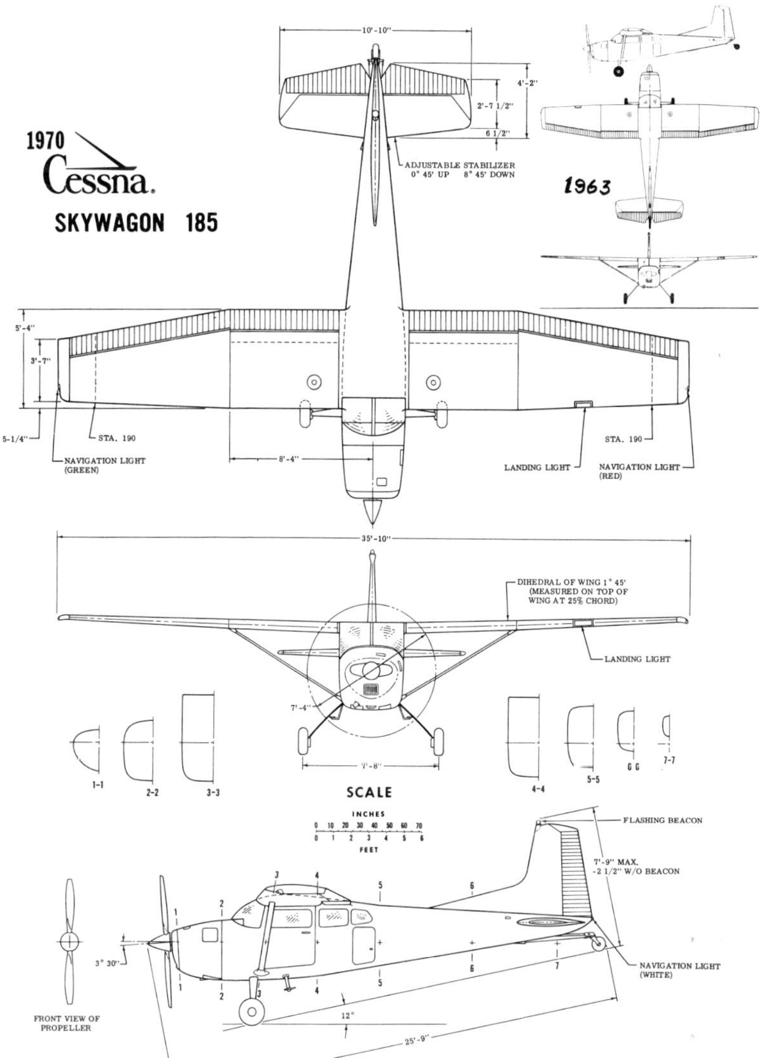
1970
Cessna
SKYWAGON 185
1963
10'-10"
4'-2"
2'-7 1/2"
6 1/2"
ADJUSTABLE STABILIZER
0° 45' UP 8° 45' DOWN
5'-4"
3'-7"
5-1/4"
STA. 190
NAVIGATION LIGHT (GREEN)
8'-4"
LANDING LIGHT
STA. 190
NAVIGATION LIGHT (RED)
35'-10"
DIHEDRAL OF WING 1° 45' (MEASURED ON TOP OF WING AT 25% CHORD)
LANDING LIGHT
7'-4"
7'-8"
1-1
2-2
3-3
4-4
5-5
6-6
7-7
SCALE
INCHES
0 10 20 30 40 50 60 70
0 1 2 3 4 5 6
FEET
FLASHING BEACON
7'-9" MAX.
-2 1/2" W/O BEACON
NAVIGATION LIGHT (WHITE)
3° 30'
12°
25'-9"
FRONT VIEW OF PROPELLER
CESSNA AIRCRAFT CO. WICHITA, KANSAS

Guide for Estimating Cost of Operation—Skywagon 185

	*PER YEAR	PER YEAR	*PER HOUR	PER HOUR	PER HOUR	PER HOUR
FLIGHT HOURS PER YEAR.				300	500	700
FLIGHT MILES PER YEAR: 168 mph, 75% power at 7,500 ft., Normal Lean Mixture.				50,400	84,000	117,600
GASOLINE: 15.9 gal./hr. at 46¢/gal.				$ 7.31	$ 7.31	$ 7.31
OIL: 1 pt./hr. plus oil change every 25 hrs. (12 qts. per change) at 63¢/qt.				$ .63	$ .63	$ .63
AIRPLANE & ENGINE MAINTENANCE.				$ 2.00	$ 2.00	$ 2.00
TOTAL OPERATIONAL COSTS (listed above).				$ 9.94	$ 9.94	$ 9.94
RESERVE FOR ENGINE OVERHAUL.				$ 2.13	$ 2.13	$ 2.13
STORAGE: at $50.00 per month.		$ 600		$ 2.00	$ 1.20	$ .86
INSURANCE (a guide for the following coverage): All risks hull coverage (Deductibles: In motion $250.00) (Not in motion $50.00).		$1,392				
Liability Coverage (Single limit $250,000) (Including passengers).		$ 140				
TOTAL INSURANCE (Industrial private business & pleasure use by a qualified pilot)		$1,532		$ 5.11	$ 3.06	$ 2.19
COST PER HOUR.				$19.18	$16.33	$15.12
COST PER MILE: 75% power, 168 mph.				$ .114	$ .097	$ .090
COST PER SEAT MILE: 6 People.				$ .019	$ .016	$ .015

FOR DEPRECIATION: Consult your tax authority. To estimate the per hour or per mile cost, divide annual depreciation amount by annual hours or miles.

*YOUR COSTS: No doubt vary from these national averages. This column is included for you to compute your estimate of your hours, miles and costs.

1970 Skywagon 185 Performance and Specifications

	LANDPLANE	AMPHIBIAN	FLOATPLANE	SKIPLANE
GROSS WEIGHT	3350 lbs. 1519 kg	*3265 lbs. 1481 kg	3320 lbs. 1506 kg	3350 lbs. 1519 kg
SPEED: Best Power Mixture				
Top Speed at Sea Level	178 mph 286 kph	156 mph 251 kph	162 mph 261 kph	157 mph 252 kph
Cruise, 75% Power at 7500′	169 mph 272 kph	149 mph 240 kph	155 mph 249 kph	150 mph 241 kph
RANGE: Normal Lean Mixture				
Cruise, 75% Power at 7500′	660 mi. 1062 km	575 mi. 925 km	600 mi. 965 km	580 mi. 933 km
62 Gallons usable fuel	3.9 hrs.	3.9 hrs.	3.9 hrs.	3.9 hrs.
No Reserve	168 mph 270 kph	148 mph 238 kph	154 mph 248 kph	149 mph 240 kph
Cruise, 75% Power at 7500′	860 mi. 1384 km	755 mi. 1215 km	785 mi. 1263 km	760 mi. 1223 km
81 Gallons usable fuel	5.1 hrs.	5.1 hrs.	5.1 hrs.	5.1 hrs.
No Reserve	168 mph 270 kph	148 mph 238 kph	154 mph 248 kph	149 mph 240 kph
Optimum Range at 10,000′	1075 mi. 1730 km	910 mi. 1464 km	970 mi. 1561 km	930 mi. 1497 km
81 Gallons usable fuel	8.3 hrs.	9.0 hrs.	9.0 hrs.	7.4 hrs.
No Reserve	129 mph 208 kph	101 mph 162 kph	108 mph 174 kph	126 mph 203 kph
RATE OF CLIMB AT SEA LEVEL	1010 fpm 308 mpm	970 fpm 296 mpm	960 fpm 293 mpm	
SERVICE CEILING	17,150 ft. 5227 m	15,300 ft. 4663 m	16,400 ft. 5000 m	
TAKEOFF:		On Land / On Water		
Ground Run (Or Water Run)	770 ft. 235 m	670 ft. 204 m / 885 ft. 270 m	1105 ft. 337 m	
Total Distance Over 50′ Obstacle	1365 ft. 416 m	1275 ft. 389 m / 1430 ft. 436 m	1740 ft. 530 m	
LANDING:				
Ground Roll (Or Water Run	480 ft. 146 m	780 ft. 238 m / 600 ft. 183 m	640 ft. 195 m	
Total Distance Over 50′ Obstacle	1400 ft. 427 m	1240 ft. 378 m / 1480 ft. 451 m	1530 ft. 466 m	
STALL SPEED:				
Flaps Up, Power Off	67 mph 108 kph	65 mph 105 kph	65 mph 105 kph	67 mph 108 kph
Flaps Down, Power Off	59 mph 95 kph	57 mph 92 kph	57 mph 92 kph	59 mph 95 kph
EMPTY WEIGHT: (Approximate) (One Seat)	1575 lbs. 714 kg	2135 lbs. 968 kg	1890 lbs. 857 kg	1725 lbs. 782 kg
USEFUL LOAD	1775 lbs. 805 kg	1130 lbs. 513 kg	1430 lbs. 649 kg	1625 lbs. 737 kg
BAGGAGE: (Total Capacity)	400 lbs. 181 kg	400 lbs. 181 kg	400 lbs. 181 kg	400 lbs. 181 kg
WING LOADING	19.3 lbs/ft^2 94.2 kg/m^2	18.8 lbs/ft^2 91.8 kg/m^2	19.1 lbs/ft^2 93.3 kg/m^2	19.3 lbs/ft^2 94.2 kg/m^2
POWER LOADING	11.2 lbs/hp 5.1 kg/hp	10.9 lbs/hp 4.9 kg/hp	11.1 lbs/hp 5.0 kg/hp	11.2 lbs/hp 5.1 kg/hp
FUEL CAPACITY: Total				
Standard	65 gal. 246 liters	65 gal. 246 liters	65 gal. 246 liters	65 gal. 246 liters
Optional Long Range Tanks	84 gal. 318 liters	84 gal. 318 liters	84 gal. 318 liters	84 gal, 318 liters
OIL CAPACITY: Total	12 qts. 11.4 liters	12 qts. 11.4 liters	12 qts. 11.4 liters	12 qts. 11.4 liters
POWER	Six-Cylinder, Fuel-Injection Engine, 300 Rated Takeoff HP (5 min) at 2850 RPM and 285 Maximum Continuous HP at 2700 RPM			
PROPELLER: Constant-Speed, Diameter	82 in. 2.08 m	86 in. 2.18 m	86 in. 2.18 m	86 in. 2.18 m
WING SPAN	35 ft 10 in 10.92 m	35 ft 10 in 10.92 m	35 ft 10 in 10.92 m	35 ft 10 in 10.92 m
WING AREA	174 $ft.^2$ 16.2 m^2	174 $ft.^2$ 16.2 m^2	174 $ft.^2$ 16.2 m^2	174 $ft.^2$ 16.2 m^2
LENGTH	25 ft 9 in 7.85 m	27 ft 6 in 8.39 m	27 ft 8.24 m	25 ft 9 in 7.85 m
HEIGHT	7 ft 9 in 2.36 m	12 ft 8 in 3.86 m	12 ft 2 in 3.71 m	7 ft 9 in 2.36 m

* A gross weight of 3265 pounds is allowed for takeoff on land only. For takeoff on water or landing on land or water, the maximum allowable gross weight is 3100 pounds (1406 kg).

The 1958 Model 172 was the third of this series and still much resembled a 170B with tricycle gear and 180-type rudder. The first 172 was the 1956 Model. Performance has changed little over the years.

CHAPTER 5: CESSNA 172/SKYHAWK AND 175/SKYLARK

The Cessna 172 and Skyhawk owner – who represents at least 10% of general aviation – has largely been misunderstood by the old hands in this business whose lives have always revolved around airplanes. Those of us who never wanted to be anything but a "pilot" from the time we were nine years old (and forty years ago that seemed to us a lofty ambition indeed), have been puzzled by all these bright young men ("newcomers") who buy and fly modern lightplanes merely for the sake of convenience.

It's the Generation Gap, Dad. The Travel Air 4000 and the grass landing patches we remember with such fondness possess no more relevance for today's young man on the way up than did the horse and buggy for us. The average Skyhawk owner probably never heard of a Gee Bee Racer; and what's more, he hasn't time to listen while we tell him about it with a faraway look in our eye. The horrible truth is, he's not interested.

Cessna recognized the emergence of this new breed and correctly assessed its attitude back in the mid-fifties. Clearly, there was a big and growing market for a four-place business/pleasure aircraft that did not demand professional-type skills and experience for its safe operation. A practical machine to serve the man who regarded a personal airplane as a kind of super-car. Cessna's supercar is the 172/Skyhawk.

The 172 was introduced in 1956, and it was, in effect, a 170 with tricycle landing gear and 180 tail. This happy combination was squarely on target; and though it has submitted to a number of changes over the year – mostly to make it prettier – the 1971 Model 172 and Skyhawk is still pretty close to the original, performance-wise, as a look at the model comparison charts (following Chapter 10) will reveal.

A brief flight evaluation of a 1957 Model 172, followed by checks of a 1962 Skyhawk and a 1968 Skyhawk should provide good representative coverage of this series. The Skyhawk, as you undoubtedly know, is simply the deluxe version of the 172, the principal differences being that the Skyhawk has a full gyro panel activated by an internal vacuum system, full paint job and speed fairings on the wheels.

The 1957 Model 172 was in excellent condition and as neat inside as TLC (Tender Loving Care) could keep it after thirteen years. Our only beef was that the flight instruments were not well grouped, but had apparently been stuck in the panel at random. An aging mechanic insisted that this was Cessna's standard arrangement for that airplane. Maybe so; we couldn't remember.

This craft sits a little higher off the ground than later models; and the old starter pull-knob is right beside the parking brake, making it easy to grab the wrong one. Otherwise, you feel right at home in this plane if you've flown later Cessnas. Firing up the engine and taxiing is standard Cessna; it rolls easily with the merest hint of power. The nosewheel's steerability is fairly limited via the rudder pedals, but it will swivel through an arc of about 30-degrees by applying full rudder and toe-brake for tight turns on the ground.

Take-off was the same now-famous "drive 'er down the runway and into the air" of later Cessnas. Then, with three aboard and full tanks (about 230 lbs below gross of 2,200 lbs), the '57 Model showed an initial climb of 500 fpm at 78 mph on a warmish day. At 6,500 ft with 2,400 rpms, which we judged to be 65% power, the IAS (Indicated Air Speed) was 106 mph. Corrected for temperature and altitude, this worked out to slightly over 120 mph TAS (True Air Speed). This latter figure was determined courtesy of Mr. Dalton's E-6B Computer; but we must confess to an attachment for the ancient (a la Travel Air days) method of arriving at this number by simply adding 2% per thousand feet of altitude to the IAS. It's not precise, but usually close enough for lightplane VFR when lapse rates are normal.

Stall characteristics of Cessna's 1957 Model Super-Car are also "Standard Cessna," which is to say you have to work at it to force this airplane into full-stall with power off. It prefers to just sit there and mush along with the stall warning horn assaulting your ears and the airspeed indicator bumping at 50 mph. It's easier to make the plane break from a stall with power on, but the air speed falls clear down to 40 mph on the dial. Some aileron control remains even at these indicated speeds, but it's probably wiser to level the wings with the rudder, especially if the flaps are down, and even more especially if you're at or near gross load. We're not sure whether it would happen with a Cessna, but we do know that sudden aileron movement in or near the stall will add just enough drag on the down-aileron side to stall-out that wing and induce a quick spin on some airplanes.

The owner of this airplane demonstrated landings holding a final approach speed of 70 mph. He touched-down on the main wheels in a full-stall from the flare, and allowed the nosewheel to come down immediately. This is the "drive it back on the ground" method and seems to work very well for a lot of people, although many instructors used to recommend holding the nose-wheel off until it eased down of its own accord. With the nose-wheel on the pavement, one does have much easier and better control, however. Also, the Cessna nose-wheel is attached to the airframe rather than the engine mount, and for short-field landings the Owner's Manual advises that the nose-wheel be put down immediately after touch-down.

The 1962 Skyhawk we checked was also well-equipped and in top condition. Except for its paint scheme, differently shaped speed fairings on the wheels and lack of a rear window, it closely resembled its newer sisters. Inside, the instrument panel held the six flight instruments in two rows directly in front of the pilot (airspeed, directional gyro and attitude gyro across the top; altimeter, turn/bank and rate-of-limb across the bottom). Engine controls were grouped at bottom-center of panel, of course, along with the fuel-sump drain knob. It's a matter of opinion, but in our view there are a few things necessary to a happy flight that should be checked *visually* and positively, like fuel tanks, oil level, tires, etc., and we just don't trust that painless (and blind) sump-drain operation from inside the cabin.

Fuel is gravity-fed to the 145-hp Continental 0-300-D engine via the fuel selector switch located between the front seats. Its

1971 Cessna Skyhawk

Nose Mounted Landing Light

Tubular Widetrack Gear

1971 Cessna Skyhawk

positions are *Off, Left, Right* and *Both,* although we get the impression that few pilots ever use fuel from the left or right wing-tank individually. Total fuel is 42 gallons, with slightly less than 41 usable.

Engine starting procedure is standard: Battery switch, "ON;" mixture set at "Full Rich," throttle slightly cracked, a couple of squirts of primer and turn ignition key past "Both" to start. We note that this owner checks his mags at 1,700 rpms, pivots his Skyhawk in a complete 360-degree circle so he can get a look at the entire pattern around this uncontrolled field, then drives down the runway and into the air, lifting off at about 60 mph without flaps. It is a cool day, about 50 degrees F at field elevation of 1,180 ft. The take-off run is estimated at 750 feet, with two of us aboard and full tanks (340 lbs below gross).

Initial climb registers 900+ fpm with maximum power and 70 mph. This drops back to 500 fpm as we lower the nose and increase speed to 100 mph with 2,500 rpms. Our host tells us that the 1962 Skyhawk's best cruising speed comes at 7,000 feet, where 2,500 rpms produces about 134 mph TAS for him. However, we stayed at 3,500 where 116 mph IAS translated into 124 mph TAS at 38 degrees F. Our friend told us that his 2,500 rpm cruise consumed 8½ gallons of fuel per hour.

Returning to his home field, our host drove his Skyhawk onto the runway holding 70 mph over the fence and let his nose-wheel come down almost immediately after the main wheels touched. He then raised his flaps and held a little back-pressure on the control wheel during roll-out. He braked sparingly into a 3-5 mph wind and used an estimated 600 feet of runway.

The 1968 Skyhawk had recently had a thorough work out at the hands of our colleague, Miss Page Shamburger. Page has about 5,000 hours in her logbook and now owns a Bonanza; but she started out in Cessnas (140's and 170's), and says she "feels right at home with these fellers." Since she spent considerable time in Skyhawk N83314L, we asked her for her impressions.

"Beginning with this 1968 Model, the 172/Skyhawk has the 150-hp 'Blue Streak' Lycoming engine," Page said, "but I really can't see where it adds anything, except maybe a little dab of useful load. This is because the 1968 Model is 40 pounds lighter than the '67 Model. Gross remains at 2,300 pounds, where it has been since 1963.

"Inside, instruments are T-grouped, and dry vacuum pumps have replaced the plumbing and oil-lubricated pumps for gyro instruments. Electrically operated flaps are standard on both the 172 and Skyhawk. The rear-vision mirror, by the way, costs $20 extra, and I found it useless, at least for rear-visioning.

"Normal engine-start procedures awaken the Lycoming's hundred and-fifty horses, and I thought this airplane seemed markedly quieter inside, probably because the twin exhaust stacks of the Continental-powered Skyhawks and 172's have been changed to a single and longer stack on the Lycoming-engined craft.

"Since this airplane was a factory demonstrator, it was loaded with avionics and extras, bringing its empty weight up to 1,407 pounds. However, I had left 893 pounds useful, which seemed enough for 'most anything. Checking my notes, I'm reminded that, taxiing out, the shorter nose-strut, introduced in 1967 to reduce in-flight drag, also allows you to see the concrete very close ahead.

"With 297 pounds of people (two hundred for Cessna's John Gallaher; ninety-seven for me), and full tanks, I was 340 pounds under gross. And according to my notes, I lifted-off in less than 800 feet at 65 mph (elevation, 1,370 ft), and remarked on the secure feeling of the Skyhawk even during those early, crucial seconds of flight. I felt that I had stability to spare. I climbed at 85 mph showing 750 fpm.

"Leveling off at 10,500 feet, I found my TAS by using the handy computer on the airspeed indicator. This is a rotating bezel mounted on the periphery of that instrument. You rotate it to set pressure ailtitude opposite outside air temperature (OAT), then read true air speed directly below the indicator needle. With an OAT of 22 degrees, I had 134 mph true at 65% power.

"During let-down for Oklahoma City, I tried some stalls and was intrigued with the pre-stall warning given by the reeds inside the wing that cry at an increasingly-higher pitch as the wing gets nearer the stall. This simple system works from air pressure flowing over the reeds, and no exterior sensor vane — so vulnerable to accidental bending — is needed. Straight ahead, with flaps up and power off, the nose sticks 'way up in the air, where the controls get heavy enough to prompt you to crank in some elevator trim (via the control located on the center console). Then this airplane broke clean at 50 mph indicated — 57 mph corrected — shaking and whistling frantically at me for mistreating it so. Flaps lower the stall to 41 IAS (49 mph corrected), and the nose-up angle is strictly unreal.

"Landings? Well, it's a Cessna; just no problem. I did the first one holding 80 mph, and flaring at that speed produced some 'ballooning.' Second time around, with half-flaps and indicating 73 mph, I put it where I wanted it and it stayed there."

The 1971 Skyhawk/172 possessed a few improvements over previous models, the main one being a switch to the Cardinal-proved tubular landing gear legs. These tapered tubular struts, encased in streamlined housings, are superior to the "slab steel" struts because they absorb shocks fore-and-aft as well as up-and-down. In our view, this is the most practical, versatile and trouble-free landing gear offered on any light airplane. We first encountered this gear back in November, 1967, when it was introduced on the original Cardinal/177. We were sold on it then. We still are. If you think the Skyhawk/172 has been easy to land in the past, this gear is, surely, the meringue on that piece of pie.

The 1971 Model also had re-styled wheel fairings. Cessna apparently re-shapes these things more for model identification than anything else. The latest ones may *look* racy, but they are undoubtedly less efficient aerodynamically than the old fashioned tear-drop shape.

Other new features for 1971 included a re-designed nose cap to eliminate propeller "slap" noise in the cabin and nose-mounted landing light. Optional, were new overhead windows, wingtip

strobe lights (white), a hands-off boom microphone and full paint job on the 172 version. The fully-articulating seats, introduced on the 1970 Model, may be adjusted to any position from fully reclined to full forward, and through a wide range of heights. The '71 Model, for the first time, offered gen-u-wine leather upholstery as an option with the Skyhawk. Most impressive of all, perhaps, was the welcome news in 1971 that the 150-hp Lycoming engine had a recommended time between overhaul of 2,000 hours. When Cessna switched to this engine in 1968, recommended TBO was 1,500 hours. It was 1,200 hours for the 0-300 Continental in earlier Skyhawk/172's.

For total utility, the re-designed instrument panel of the 1971 Model offers plenty of space for magical black boxes, including 300 Nav-O-Matic autopilot, 300 Transceiver, 90 and 360-channel Nav/Coms, ADF, Marker Beacon, Glideslope Receiver, the 300 Transponder with a reply code capability of up to 4,096 with Modes A and AC, and the ARC (Cessna) DME.

Performance of the 1971 Models were not much changed, as a check of the Model Comparison Tables at the back of this book will confirm. List price of the basic 172 for 1971 was $13,425, and the '71 Skyhawk list was $14,995.

The Army and Air Force Cessna T-41 is the military version of the 172, and Cessna records showed that about 500 T-41's had been delivered as of October 1, 1970.

The Cessna 175/Skylark was brought out in 1958 and, for a couple of years, sold almost as well as the 172/Skyhawk (the Skyhawk version of the 172 first appeared in 1959); but the 175 slipped badly in sales in 1961, and production ended in 1963, at which time Cessna briefly called this airplane the "172 Powermatic."

The 175/Skylark and Powermatic is basically a 172 airframe fitted with the Continental G0-300-E engine of 175 hp, and constant speed propeller. The G0-300-E is a geared mill which produces its maximum power at 2,400 propeller rpms while the crankshaft rpms register 3,200, a ratio of .75:1. Since the propeller reduction-gear box is located above the drive shaft on the front of the engine, the same 84-inch propeller as used on the 145-hp engines gains a couple of inches in ground clearance. And since this is the same engine as the 145-hp 0-300-D employed by Skyhawks of this period – except for the gearing – it's obvious that the additional 30 horsepower results from the higher crankshaft speed, and *that* clearly can't *add* anything to engine life or reliability.

This model has an extra knob in the cockpit: the prop pitch control. It is also a push-pull type and changes the setting of the propeller governor to control engine speed. This control may be moved through its full range by depressing a locking button in the center of the knob, while minor adjustments are made by releasing the locking button and rotating the knob – clockwise to increase rpms and counter-clockwise to decrease rpms.

For all ground operations and take-off, the propeller control should be full in (high rpms). After take-off, reduce throttle first, then reduce rpms. Since a small control movement will produce a considerable rpm change, you should set up climb and cruise rpms by screwing the knob in or out.

This airplane also has cowl flaps which are not found on comparable models of the 172. These of course merely control the flow of air over the engine for greater efficiency under varying conditions. Opening the cowl flaps on the ground and during steep climbs, for example, improves engine cooling. In flight, closing the flaps reduces the cooling air to the engine and reduces drag. This control is a double-button type with a friction lock.

U.S. Army (Cessna) T-41B

1971 Cessna 172L AND SKYHAWK

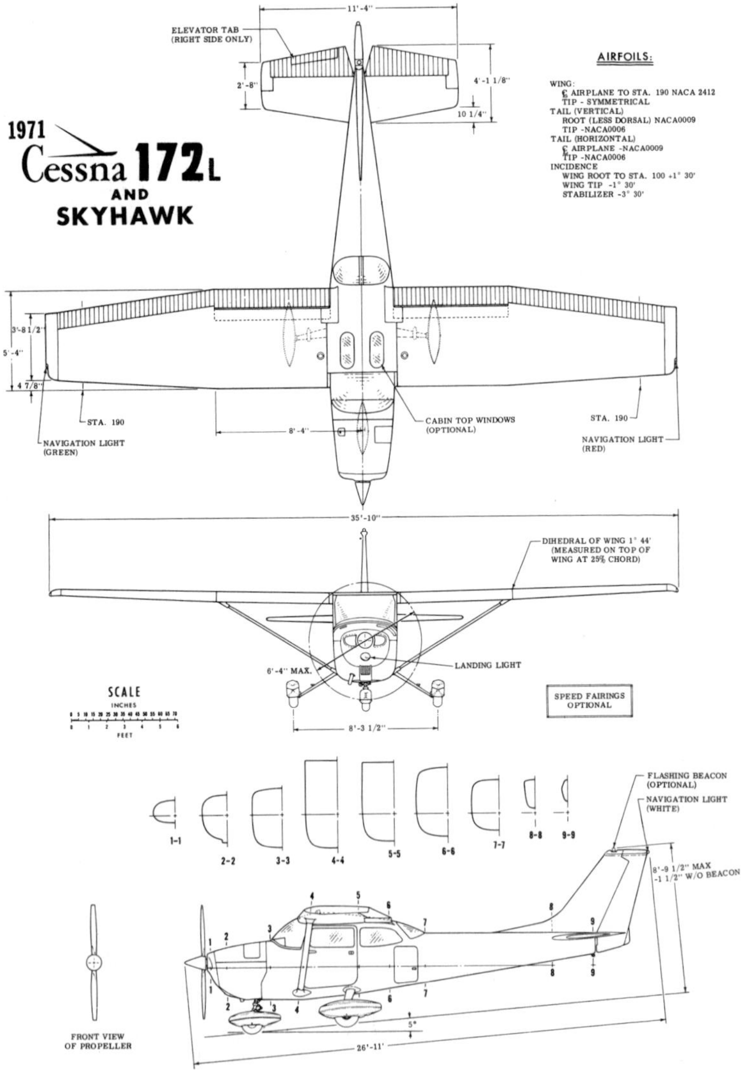

These planes use the same fuel system as that of the 172/Skyhawk, except that capacity is 10 gallons greater.

This series of course has improved performance compared to the standard 172's, but the difference was probably not great enough to warrant the extra cost. The 1958 Cessna 175 had a basic list of \$10,995 vs. \$8,995 for the 172. By 1963, the "Powermatic 172" was priced at \$13,275 vs. \$10,245 for the 172, while the 1963 "Skyhawk Powermatic" sold for \$14,650 vs. \$11,995 for the 1963 145-hp Skyhawk.

Cessna 175/Skylark and Powermatic specifications and performance figures follow at the end of this chapter.

SPECIFICATIONS, CESSNA 175/SKYLARK AND POWERMATIC

	175 (1958)	Powermatic (1963)
Engine	Continental G0-300-E	Continental G0-300-E
Wingspan	36 ft 2 in	36 ft 2 in
Length	25 ft	26 ft 6 in
Height	8 ft 6 in	8 ft 11 in
Wing Area	174 sq ft	174 sq ft
Gross Weight	2,350 lbs	2,500 lbs
Wing Loading	13.5 lbs sq ft	14.4 lbs sq ft
Power Loading	13.4 lbs/hp	14.3 lbs/hp
Useful Load	945 lbs	1,140 lbs
Fuel Capacity	52 gals	52 gals

PERFORMANCE, CESSNA 175/SKYLARK AND POWERMATIC

	175 (1958)	Powermatic (1963)
Take-Off	640 ft	600 ft
clear 50 ft	1,340 ft	1,205 ft
Landing	600 ft	610 ft
clear 50 ft	1,155 ft	1,200 ft
Climb, Initial	850 fpm	830 fpm
Service Ceiling	17,800 ft	17,000 ft
Maximum Speed	147 mph	146 mph
Cruise (75%)	140 mph	138 mph
Range (@75%)	585 mi	540 mi
Stall (no flaps)	58 mph	64 mph
Stall (flaps 40-deg)	50 mph	55 mph

1971 Model Skyhawk/172 Performance and Specifications*

	MODEL 172		SKYHAWK		FLOATPLANE	
GROSS WEIGHT	2300 lbs.	1043 kg	2300 lbs.	1043 kg	2220 lbs.	1007 kg
SPEED:						
Top Speed at Sea Level	139 mph	224 km/h	140 mph	225 km/h	108 mph	174 km/h
Cruise, 75% Power at 9000'**	131 mph	211 km/h	132 mph	212 km/h	106 mph	171 km/h
RANGE:						
Cruise, 75% Power at 9000'**	615 mi.	990 km	620 mi.	995 km	500 mi.	805 km
38 Gallons, No Reserve	4.7 hr.	4.7 hr.	4.7 hr.	4.7 hr.	4.7 hr.	4.7 hr.
	131 mph	211 km/h	132 mph	212 km/h	106 mph	171 km/h
Cruise, 75% Power at 9000'**	775 mi.	1247 km	780 mi.	1255 km	625 mi.	1005 km
48 Gallons, No Reserve	5.9 hr.	5.9 hr.	5.9 hr.	5.9 hr.	5.9 hr.	5.9 hr.
	131 mph	211 km/h	132 mph	212 km/h	106 mph	171 km/h
Optimum Range at 10,000'	640 mi.	1030 km	655 mi.	1050 km	530 mi.	855 km
38 Gallons, No Reserve	5.5 hr.	5.5 hr.	5.5 hr.	5.5 hr.	5.5 hr.	5.5 hr.
	117 mph	188 km/h	118 mph	190 km/h	97 mph	156 km/h
Optimum Range at 10,000'	820 mi.	1320 km	830 mi.	1336 km	670 mi.	1075 km
48 Gallons, No Reserve	7.0 hr.	7.0 hr.	7.0 hr.	7.0 hr.	7.0 hr.	7.0 hr.
	117 mph	188 km/h	118 mph	190 km/h	97 mph	156 km/h
RATE OF CLIMB AT SEA LEVEL	645 fpm	196 m/m	645 fpm	196 m/m	580 fpm	177 m/m
SERVICE CEILING	13,100 ft.	3993 m	13,100 ft.	3993 m	12,000 ft.	3658 m
TAKEOFF: Ground Run (or Water Run)	865 ft.	264 m	865 ft.	264 m	1620 ft.	494 m
Total Distance Over 50-ft Obstacle	1525 ft.	465 m	1525 ft.	465 m	2390 ft.	729 m
LANDING: Ground Roll (or Water Run)	520 ft.	159 m	520 ft.	159 m	590 ft.	180 m
Total Distance Over 50-ft Obstacle	1250 ft.	381 m	1250 ft.	381 m	1345 ft.	410 m
STALL SPEED: Flaps Up, Power Off	57 mph	92 km/h	57 mph	92 km/h	59 mph	95 km/h
Flaps Down, Power Off	49 mph	79 km/h	49 mph	79 km/h	52 mph	84 km/h
EMPTY WEIGHT (approximate)	1250 lbs.	567 kg	1300 lbs.	590 kg	1415 lbs.	642 kg
USEFUL LOAD	1050 lbs.	476 kg	1000 lbs.	453 kg	805 lbs.	365 kg
BAGGAGE	120 lbs.	54 kg	120 lbs.	54 kg	120 lbs.	54 kg
WING LOADING	13.2 lbs/ft^2	64 kg/m^2	13.2 lbs/ft^2	64 kg/m^2	12.7 lbs/ft^2	62 kg/m^2
POWER LOADING	15.3 lbs/hp	7.0 kg/hp	15.3 lbs/hp	7.0 kg/hp	14.8 lbs/hp	6.7 kg/hp
FUEL CAPACITY: Total						
Standard Tanks	42 gal.	159 liters	42 gal.	159 liters	42 gal.	159 liters
Optional Long Range Tanks	52 gal.	197 liters	52 gal.	197 liters	52 gal.	197 liters
OIL CAPACITY	8 qts.	7.6 liters	8 qts.	7.6 liters	8 qts.	7.6 liters
POWER: Four-Cylinder Engine	150 rated hp at 2700 rpm					
PROPELLER: All-Metal Fixed-Pitch Dia.	76 in.	1.93 m	76 in.	1.93 m	80 in.	2.03 m
WING SPAN	35 ft. 9 1/2 in.	10.91 m	35 ft. 9 1/2 in.	10.91 m	35' 9 1/2 in.	10.91 m
WING AREA	174 sq. ft.	16.16 m^2	174 sq. ft.	16.16 m^2	174 sq. ft.	16.16 m^2
LENGTH	26 ft. 11 in.	8.21 m	26 ft. 11 in.	8.21 m	27 ft.	8.23 m
HEIGHT (With Depressed Nose Strut on Landplane)	8 ft. 9 1/2 in.	2.68 m	8 ft. 9 1/2 in.	2.68 m	9 ft. 11 in.	3.02 m

*Subject to change without notice
**At 6500 ft. for Floatplane

Guide for Estimating Cost of Operation—Skyhawk & 172

	*PER YEAR	PER YEAR	*PER HOUR	PER HOUR	PER HOUR	PER HOUR
FLIGHT HOURS PER YEAR.				300	500	700
FLIGHT MILES PER YEAR: 132 mph, 75% power at 9,000 ft., Lean Mixture.				39,600	66,000	92,400
GASOLINE: 8.1 gal./hr. at 43¢/gal.				$ 3.48	$ 3.48	$ 3.48
OIL: 1pt/hr. plus oil change every 50 hrs. (8 qts. per change) at 63¢/qt.				$ .42	$ .42	$ .42
AIRPLANE & ENGINE MAINTENANCE.				$ 1.50	$ 1.50	$ 1.50
TOTAL OPERATIONAL COSTS (listed above).				$ 5.40	$ 5.40	$ 5.40
RESERVE FOR ENGINE OVERHAUL.				$ .81	$ .81	$ .81
STORAGE: at $50.00 per month.		$ 600		$ 2.00	$ 1.20	$ .86
INSURANCE (a guide for the following coverage): All risks hull coverage (Deductibles: In motion $250.00) (Not in motion $50.00).		$1,026.				
Liability Coverage (Single limit $250,000) (Including passengers).		$ 140				
TOTAL INSURANCE (Industrial private business & pleasure use by a qualified pilot).		$1,166		$ 3.89	$ 2.33	$ 1.67
COST PER HOUR.				$12.10	$ 9.74	$ 8.74
COST PER MILE: 75% power, 132 mph.				$ .092	$ .074	$ .066
COST PER SEAT MILE: 4 People.				$ .023	$ .019	$ .017

FOR DEPRECIATION: Consult your tax authority. To estimate the per hour or per mile cost, divide annual depreciation amount by annual hours or miles.

*YOUR COSTS: No doubt vary from these national averages. This column is included for you to compute your estimate of your hours, miles and costs.

1968 Cessna Cardinal

CHAPTER 6: THE CESSNA 177/CARDINAL

When the first 177 and Cardinal was introduced in November, 1967, with a swinging party for dealers and av-writers at Cessna's Wichita Delivery Center, we were there for a serious look at this all-new design. While most of our colleagues were swapping lies, enjoying the Dixieland band and stuffing themselves with bar-b-que, we slipped off with a company check-pilot for an afternoon of air work in 2207Y, one of the first Cardinals built (1968 Model).

We liked that airplane, and found almost nothing to pick at. However, owners soon discovered to their horror that it didn't fly or land exactly like a strutless Skyhawk, and some heavy-handed Super-Car drivers managed to smash the Cardinal tail into the pavement on landing, knock-off a few nose-wheels, etc. (apparently, this was possible if one closed his eyes, used full back-pressure on the wheel at the flare and then sat rigidly waiting for the crashing noises to subside).

Cessna accepted the blame gracefully. That was proper because, after all, they had lulled a generation of Cessna pilots into near-effortless flying with the extremely-forgiving 172 series, and it probably wasn't ethical to suddenly offer, to many of those same customers, a Cessna that didn't handle exactly like a Cessna. The company therefore picked up the tab for a list of modifications that gentled the Cardinal and returned the smiles to the faces of dealers and customers.

The principal modification made to the first Cardinals was installation of inverted slots near the leading edge of the stabilator, because it was discovered that, in a full nose-up attitude, the stabilator was stalling out. This was Cessna's first application of a horizontal stabilator (which replaces the more conventional horizontal stabilizer and elevators), and it was chosen for the Cardinal because, with the wing placed so far back for improved pilot visibility and better balance, tail surfaces would have had to be considerably enlarged or the fuselage lengthened to achieve adequate control using a stabilizer and elevator system. The stabilator is more effective for its size and saves both weight and drag.

When the Cardinal first appeared, a Cessna engineer told us that it represented an engineering attempt to design the airplane of the seventies; a plane that would set the trend at Cessna for the coming decade. That was in the fall of 1967; and two years later, when the 1970 Model was announced, we believed that the Cardinal had evolved into that very thing. The 1969 Model went to 180-hp (from the original 150-hp); then the 1970 Model added a constant-speed propeller to efficiently employ those 30 extra horses, and *also* received an all new wing (a new airfoil shape), a wing that at last made the sleek Cardinal "fly like a Cessna."

Then, obviously feeling that they had combined all the right things in a single airframe, Cessna coasted with the 1971 Model Cardinal and 177, offering no other significant changes except a re-designed nose-plate containing the landing light, plus some minor interior pretties.

About 1,000 of the original 1968 Models (150-hp) were built, and slightly over 400 of the 180-hp 1969 Cardinals followed. The real story of the Cardinal, however, will be told during the early seventies, with the new wing, constant-speed propeller and tamed stabilator control combining a "Cessna ride" with increased performance – and offered in an expanding market (following the recession of 1970).

Going back to the tapes we made when we were introduced to the first Cardinal, we're reminded that that caper was a magazine assignment and, looking for a fresh angle to interest the reader, we invited a 20 year-old girl pilot, Miss Saundra Nix, to go along and do most of the flying so she could give her impressions. Sandy was a sophomore at Oklahoma State University at the time, but the holder of a Flight Instructor's Rating. She had soloed at 16 and obtained her Commercial License at 18.

Sandy had been instructing in 150's, and was enchanted with the Cardinal's sports car look, wide wheel-tread and its soft ground ride (this was Cessna's first use of the shot-peened vanadium-steel tubular gear legs). The plane's cabin was "huge" in Sandy's estimation, and so comfortable it was "unreal."

With Sandy flying and the check-pilot in the right-front seat, we had the spacious back seat to ourselves. So, we kept the tape recorder going and took notes on instrument readings in all flight regimes. Lift-off was normal, requiring eleven seconds and about 900 feet. We had a 3-5 knot south wind, ground temp of 80 degrees F, and were approximately 300 lbs under allowable gross. Sandy stayed low until we were clear of the McConnell AFB jet pattern, then climbed at 90 mph which resulted in 800 fpm, dropping to about 700 fpm as we passed through 4,000 feet. There, she tried some stalls, which the Cardinal resisted with a noticable buffet, sort of like a V-tail Bonanza, but recovery was quick and sure and our pilot (pilotess?) reported that the airplane "felt good" throughout the stall. She also remarked that the variation between indicated air speed and calibrated air speed at high angles-of-attack seemed very small. This machine was supposed to stall at 64 mph without power and no flaps, and that's about what it indicated instead of 55 or other low figure.

Cruise, at 75% power at 9,000 feet, was listed in the pilot's

manual at 134 mph, and Zero Seven Yankee gave us 135 mph TAS under those conditions. Sandy remarked upon the effectiveness of the stabilator and fiddled with the trim, adding that this airplane "doesn't feel much like a 172."

At this point, we were approaching Eldorado, Kansas, so we went down to try some landings. Sandy was over the fence indicating 85 mph and she got a fair bounce at touch-down. She looked at the check-pilot as if he'd betrayed her. "I *never* bounce a Cessna with that old slab-steel gear on it," she accused.

"You over-controlled a little," he replied. "Most people do until they get used to the new stabilator."

"Yecch!" Sandy said, sticking out her tongue. She fed-in power and went around for another try.

After her third landing – by then coming over the fence with 80 mph indicated – Sandy was putting the Cardinal down very nicely, so she abandoned the concrete runway and began shooting landings cross-wind in the grass. The Cardinal's low profile proved very useful as well as pretty in this activity, and our pilot reported that it handled a side-wind almost as well as a low-winger. And it was during the landings and take-offs on the sod that we decided that Cessna should go to the new tubular-steel landing gear on all their airplanes. That ground was anything but smooth; but we got a smooth ground roll nevertheless, while the gear legs danced frantically in all directions. Truly, that new Cessna gear is the greatest thing since varnished propellers.

It was almost dark when we returned to Cessna Field, and

by then Sandy was much taken with the Cardinal. Climbing into her little Cessna 150, she looked back at the shiny, low-slung Cardinal and sighed, "Now I know how Cinderella felt when her lovely coach turned back into a pumpkin."

In our view, the original Cardinal wasn't as unforgiving as some have claimed. If you were paying attention and flew it by feel instead of by rote, it behaved exactly as it was supposed to. What it failed to do was respond perfectly to Skyhawk technique, because it simply was not just a strutless Skyhawk. We know a fellow out in Amarillo, Texas who owns an early Cardinal and likes it. He says he's at a loss to explain why its original owner sold it at a bargain. We think the explanation may lie in the fact that our Texas friend had not been flying a Cessna when he bought his Cardinal.

Anyway, the 1970 and 1971 models of the Cardinal and 177 (the latter being the economy version of the Cardinal) have that new wing and a domesticated stabilator (modified linkage to the control wheel), and these changes, along with the extra power and constant-speed prop, do make this a different airplane – as Cessna says, a good airplane made great.

The 1971 model is a slightly fancied-up version of the '70 model with no significant aerodynamic change. The interior was re-styled and padding added to door posts, front seat backs and lower instrument panel. The wheel pants were re-shaped again, which appears to be Cessna's way of keeping up with Detroit, and the nose-plate re-designed to contain the landing light. All the good things were kept from previous models, of course, including the

1971 Cessna Cardinal

four-foot-wide doors which provide the easiest cabin entry and exit in the industry, and the 115 cubic-foot cabin remains the biggest in its class, offering posh surroundings for four adults and two children, or four people and lots of luggage. The 1971 Cardinal had a list of $17,995 while the Model 177 was priced at $16,795 fly-away-factory, Wichita.

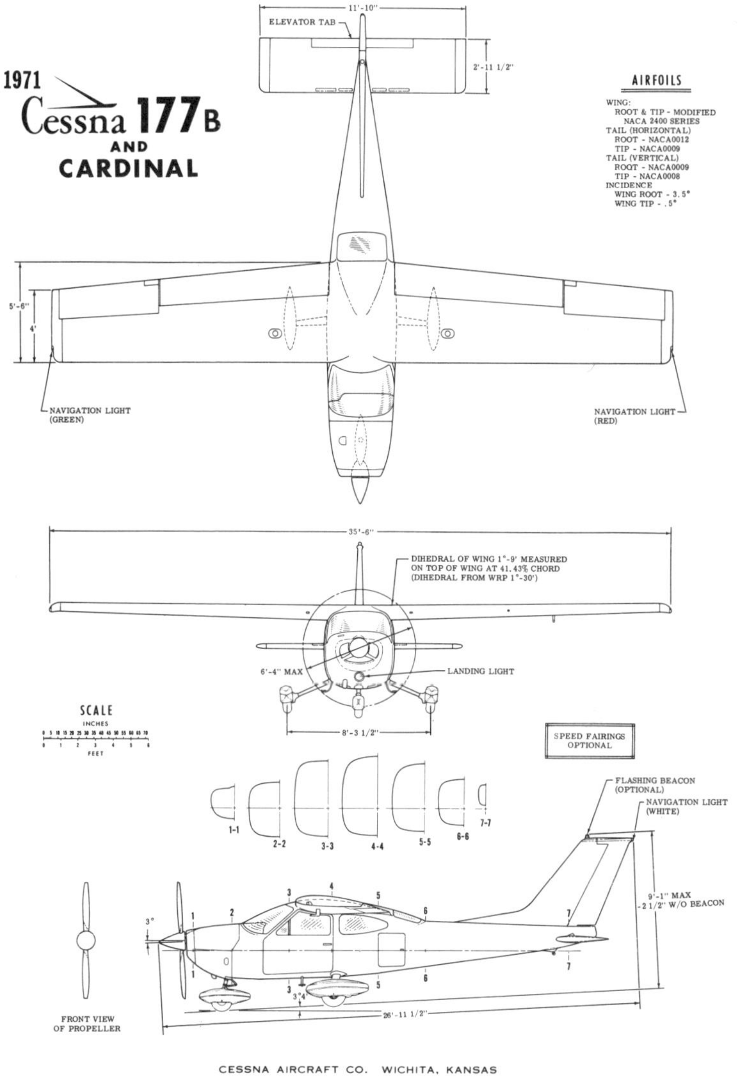

1971 Model Cardinal and 177 Performance and Specifications*

	177		CARDINAL	
GROSS WEIGHT	2500 lbs.	1134 kg	2500 lbs.	1134 kg
SPEED: Top Speed at Sea Level	150 mph	241 kph	153 mph	246 kph
Cruise, 75% Power at 8000'	139 mph	224 kph	142 mph	229 kph
RANGE: Cruise, 75% Power at 8000' 49 Gallons, No Reserve	675 mi. 4.9 hr. 139 mph	1086 km 4.9 hr. 224 kph	690 mi. 4.9 hr. 142 mph	1110 km 4.9 hr. 229 kph
Optimum Range at 10,000' 49 Gallons, No Reserve	775 mi. 6.3 hr. 123 mph	1247 km 6.3 hr. 198 kph	790 mi. 6.3 hr. 125 mph	1271 km 6.3 hr. 201 kph
RATE OF CLIMB AT SEA LEVEL	840 fpm	256 mpm	840 fpm	256 mpm
SERVICE CEILING	14,600 ft.	4450 m	14,600 ft.	4450 m
TAKEOFF:				
Ground Run	750 ft.	229 m	750 ft.	229 m
Total Distance Over 50'-Obstacle	1400 ft.	427 m	1400 ft.	427 m
LANDING:				
Ground Roll	600 ft.	183 m	600 ft.	183 m
Total Distance Over 50'-Obstacle	1220 ft.	372 m	1220 ft.	372 m
STALL SPEED: Flaps Up, Power Off	63 mph	101 kph	63 mph	101 kph
Flaps Down, Power Off	53 mph	85 kph	53 mph	85 kph
EMPTY WEIGHT: (Approximate)	1430 lbs.	649 kg	1480 lbs.	671 kg
USEFUL LOAD	1070 lbs.	485 kg	1020 lbs.	463 kg
BAGGAGE	120 lbs.	54 kg	120 lbs.	54 kg
WING LOADING	14.4 lbs/ft^2	70.2 kg/m^2	14.4 lbs/ft^2	70.2 kg/m^2
POWER LOADING	13.9 lbs/hp	6.3 kg/hp	13.9 lbs/hp	6.3 kg/hp
FUEL CAPACITY	50 gallons	189 liters	50 gallons	189 liters
OIL CAPACITY	8 quarts	7.6 liters	8 quarts	7.6 liters
POWER: Four-Cylinder Engine	180 rated hp at 2700 rpm			
PROPELLER: Constant-Speed, Dia.	76 in.	1.93 m	76 in.	1.93 m
WING SPAN	35 ft. 6 in.	10.82 m	35 ft. 6 in.	10.82 m
WING AREA	174 sq. ft.	16.16 m^2	174 sq. ft.	16.16 m^2
LENGTH	26 ft. 11 1/2 in.	8.22 m	26 ft. 11 1/2 in.	8.22 m
HEIGHT: (With Depressed Nose Strut)	9 ft. 1 in.	2.77 m	9 ft. 1 in.	2.77 m

* Subject to change without notice

Guide for Estimating Cost of Operation—Cardinal & 177

	*PER YEAR	PER YEAR	*PER HOUR	PER HOUR	PER HOUR	PER HOUR
FLIGHT HOURS PER YEAR.				300	500	700
FLIGHT MILES PER YEAR: 142 mph, 75% power at 8,000 ft., Lean Mixture.				42,600	71,000	99,400
GASOLINE: 10.0 gal./hr. at 46¢/gal.				$ 4.60	$ 4.60	$ 4.60
OIL: 1 pt./hr. plus oil change every 50 hrs. (8 qts. per change) at 63¢/qt.				$ .42	$ .42	$ .42
AIRPLANE & ENGINE MAINTENANCE.				$ 1.75	$ 1.75	$ 1.75
TOTAL OPERATIONAL COSTS (listed above).				$ 6.77	$ 6.77	$ 6.77
RESERVE FOR ENGINE OVERHAUL.				$ 1.01	$ 1.01	$ 1.01
STORAGE: at $50.00 per month.		$ 600		$ 2.00	$ 1.20	$ .86
INSURANCE (a guide for the following coverage): All risks hull coverage (Deductibles $250.00 in motion) (Not in motion $50.00). . . . . **		$1,196				
Liability Coverage (Single limit $250,000) (Including passengers).		$ 140				
TOTAL INSURANCE (Industrial private business & pleasure use by a qualified pilot).		$1,336		$ 4.45	$ 2.67	$ 1.91
COST PER HOUR.				$14.23	$11.65	$10.55
COST PER MILE: 75% power, 142 mph.				$ .100	$ .082	$ .074
COST PER SEAT MILE: 4 People.				$ .025	$ .021	$ .019

FOR DEPRECIATION: Consult your tax authority. To estimate the per hour or per mile cost, divide annual depreciation amount by annual hours or miles.

*YOUR COSTS: No doubt vary from these national averages. This column is included for you to compute your estimate of your hours, miles and costs.

**Average for Cardinal.

CHAPTER 7: CESSNA 150 AND 150 AEROBAT

On February 23, 1951 the last Cessna 140A left the assembly line in Wichita, and for seven years Cessna did not offer a two-place airplane. By 1958, however, a good volume-production two-place trainer was so much in demand that flight operators were paying more for 7-year-old 140's than those craft sold for new. True, there were a couple of good fabric-covered two-placers in the market, but the economical all-metal 140A had spoiled a lot of flight operators (and customers); nothing less than a 140 follow-on would dc.

Happily, Cessna listened to those voices from the aeronautical wilderness – and produced a 140 follow-on. They called it the Cessna 150.

The first 150 appeared late in 1958, and it did not pretend to improve upon the 140's performance figures, but instead used its ten extra horsepower to do much the same with greater ease and comfort without significantly increasing cost of operation. In short, it was a scaled-down 172, and except for the pleasant fact that the new 150 had a slightly quicker control-response — seemed a bit more "lighthearted" — it inherited all of the 172's amiable characteristics. It came close to being the ideal sport/trainer, as its apparently ageless popularity has attested ever since.

Now, before we go any further we must acknowledge that a lot of people may protest the above statement. Some will maintain that the ideal sport/trainer is the Cherokeke 140; others will say it's the Musketeer Sport or the Citabria, etc. Therefore, we've got to hedge by admitting that it all depends upon who's talking. It isn't possible to (honestly) say that one airplane is "better" than another, because *every* aircraft design is a compromise; to get something, you have to give something. If you want speed, for example, you must pay for it with other desirable traits, including economy of operation. So, the most anyone can do is decide which airplane is best for him; and that, clearly, is determined by each individual's preferences, prejudices and, perhaps, the personal weaknesses most of us possess. But we can allow our "ideal" statement to stand if we add "for the most people," because the Cessna 150 has far out-sold all other planes in its class.

In 1969, 61% of all flight-training hours were flown in Cessnas, primarily the 150. Of the 6,800,000 flight-training hours flown that year in the U.S. by all types of aircraft, 3,100,000 flight-training hours were accounted for by the Cessna 150. So, who can argue with success of that magnitude?

The 150, like most airplanes, has grown a little heavier as the years pass, and has paid for it with a little bit of performance. This seems inevitable, because customers want ever more comfort and convenience, and appear willing to give up a little range/speed/climb-rate to pamper the flesh. A look at the Model Comparison Tables following Chapter 10 will show this trend, which Cessna successfully curbed after 1966. The 1962 and 1963 Models (Commuter version) had low empty weights, and slightly better performance. The big rear window was added in 1964; doors were enlarged and squared-off and the swept-rudder added in 1966. The conical-camber wingtips came with the 1970 Model, and in 1971 the 150 received that lovely tubular-steel landing gear, new noseplate with landing light and extended prop shaft, plus new dorsal fin — and of course, re-shaped wheel pants (called "speed fairings" by Cessna advertising men, though they add little speed).

The conical wingtips, which improve turn and slow-flight

1971 Cessna Model 150

stability, are about the only significant aerodynamic changes in the 150 design since its inception. All the other many improvements have made it quieter, easier to handle on the ground, handier to operate in the air, better to see out of and prettier to look at. Come to think of it, that's about all Detroit has done for cars during the past thirty years.

The first 150, the 1959 Model, is still a nice little flying machine if it has been properly maintained (we probably should speak of all Cessna models in present tense, since all – back through the pre-WW-II Airmaster – are still flying in surprising numbers). The 1959 Model has the pull-starter (since replaced by the key ignition type), and a cabin more closely resembling the 140 than that of later 150's; but it flies just like its fancier off-spring of later years.

One of these early ones, say, 1959-1963, which may be found in the used plane market at prices ranging from $4,000 to $6,000, is probably a much better buy than the cheaper, fabric-covered used planes for the median-income pilot who flies mostly for fun. The 150 will not depreciate much, and its maintenance will be less costly, particularly if one is faced with a re-cover job on the rag-winger. Also, the 150 is always good merchandise in the used market when one is ready to trade or sell. During the late sixties, it was not at all uncommon for a man to buy a used 150, fly it for a couple of years, then sell it for as much as he paid for it in the first place. His cost of ownership was represented by whatever it cost him to keep it in top condition. And since the average 150 (all models) will burn only 5 to 6 gallons of fuel per hour (at 115-120 mph) it gives compact-car gas milage.

But "practical" reasons aside, the 150 also has a sort of friendly air about it that just naturally attracts people. In contrast to the rather ominous stance of the bigger, faster and more powerful craft that seems to *challenge* you to flight, the saucy little 150 *invites* you with a twinkle, as if to say, "C'mon, let's go have some fun!"

This characteristic implies, of course, that it won't place any great demands upon you. And it doesn't. We taped a demonstration ride in a 1966 Model – that was the year the swept rudder and bigger doors were added, along with larger tire size (same as 172) – and this airplane, with a little over 1,100 hours on its recording tach, still looked new late in 1970.

We were surprised to discover that the old pull-type starter lever was still used in the 150 as late as 1966, and the flap indicator, above the left window, was awkward to read; but electric flaps began with this model and the old "axe handle" flap lever disappeared from between the seats in favor of a switch on the panel. Stabilizer trim is on a console just below the throttle.

To start the engine, mixture is set at "Rich," fuel selector switch, "On" (there are no Left and Right positions for individual fuel tanks), crack throttle 1/4th of an inch, a couple of shots of "prime," clear the prop and firmly pull the starter T-handle.

As the oil pressure comes up, we ask Ground Control at Lawton (Oklahoma) Municipal for taxi instructions and then move out. She rolls easily, the 100-hp Continental muttering happily under its

1971 Cessna Model 150

breath. Nose-wheel steering with the rudder pedals is quick and positive. On the taxi strip, we set altimeter (field elevation, 1,108 ft), and determine that we are about 170 lbs below our allowable gross weight.

Prior to take-off, the magneto checks are made at 1,700 rpms. The Owner's Manual discourages full-power run-ups unless a malfunction is suspected. Besides, with plenty of runway available, the full-power check is more practical during the early part of the take-off run, because there's plenty of room to shut-down everything and stop if a problem is suspected. Completing our pre-take-off check, we make sure the doors are latched; all flight controls free; adjust stabilizer trim to "Take-Off;" check operation of carburetor heat (leaving it at "Cold"), and check flaps in "Up" position. Then, with tower's blessing, we feed-in power and accelerate down the runway centerline.

As the air speed passes 40 mph, we begin to ease back on the control wheel, and a few seconds later, as the airspeed needle passes 50 mph, the 150 "slips the surly bonds of earth" and returns to its natural habitat. We leave the engine at full power and adjust the speed at 85 mph with the control wheel. This results in an initial climb-rate of 650 feet per minute (fpm) with an OAT of 64 degrees F. A steeper angle, which brings the airspeed needle back to 75 mph, gives us 740 fpm. We leave the pattern, and continue upward on a westerly heading to get away from military and airline traffic in the area.

We lean mixture as we climb in order to maintain best rpms, and level off at 5,000 feet, which puts us well above the haze layer in clear air. A check on cruising speeds at this altitude shows 121 mph TAS turning 2,750 rpms. It drops back to about 108 mph TAS at 2,500 rpms. This latter setting should lower fuel consumption to about 5 gph (from more than 6 gph) and increase range by approximately 75 miles. It is also, of course, easier on the engine. Optimum cruise at 2,650 rpms produces 115 mph true at this height.

We try some stalls, and with power off, no flaps, the 150 mushes along, control wheel back, wings level at 48 mph IAS. With flaps down and power off, there is a more definite break at about 42 mph IAS, preceded by some buffeting, but again wings remain level and the airplane is flying almost as soon as back-pressure on the wheel is relaxed. It's a little more fun stalling out of a 30-degree bank with flaps. The break is quite clean from this attitude and the

NOSE MOUNTED LANDING LIGHT

buffeting pronounced, while the pay-off comes at about 58 mph. However, the down-wing comes up at once and recovery is swift.

The only surprise one is likely to encounter doing stalls in the later model Cessnas (beginning with the 1966 Model 150), is the wierd sound (our colleague Don Downie likened it to the "dying gasp of a collapsing bag pipe") of the Cessna pneumatic reed stall-warning device. This system, maintenance-free and non-electrical, admits the relative wind through a 3/4-inch opening in the wing's leading edge which flows through the reed device to produce an increasingly eerie sound as one aproaches the stall. It really raises the hair on the back of your neck, and should be the most effective stall-warner in the business, next to a short-tempered flight instructor.

Landings are typical Cessna landings. You can make the 150 float and porpoise coming in with too much speed and a high flare; but it'll eventually mush onto the runway and land itself anyway. Or, you can drag over the fence at 60 mph and plop it down firmly on all three wheels. It won't complain. But if you're willing to work at your landings a little, this airplane will really make you look good in the maneuver that has traditionally been the primary measuring stick of pilot proficiency.

From the time the first 150 appeared late in 1958, it has been offered in three versions, the Standard, Trainer and Commuter. The 150 Aerobat was added in 1970, and about 300 were sold during its first year in production (Still another version is the French-built F150, sold throughout Europe since 1966.)

The 1971 Model 150's were announced in September, 1970, and the most significant improvement (in our opinion) was the switch to the Cardinal-type hollow, tubular-steel gear legs. This system gives the 150 an even wider track – nearly 8 ft – and a ground ride that surely can't be bettered short of the hovercraft principle. If that isn't enough, Cessna says that this gear has four to five times the service life of previous landing gears.

Interiors were again poshed-up for '71, and the contoured cabin ceiling of easy-to-clean, rip-resistant Ensolite (a 1970 innovation) added 2½ inches to headroom. A new level of quietness resulted from the propeller extention and new cowling. There was also a new safety belt system.

List prices for the 1971 Model 150's were: Standard, $8,895; Trainer, $10,775; Commuter, $11,995 and Aerobat $10,995.

The Cessna 150 Aerobat is the versatile one. It is a trainer, cross-country personal/business plane, and a fun machine approved for all the aerobatic maneuvers most private pilots are likely to master. It is certified for barrel rolls, aileron rolls, single snap rolls, loops, Immelmann turns, spins, Cuban-8's and vertical reversements. It is stressed for six G's positive and three G's negative flight loads. Special equipment includes quick-release door mechanisms, seats with removable bottom and back cushions to accommodate either back or seat-pack parachutes, quick-release lap belts and shoulder

1971 Cessna 150 Aerobat

1971 Cessna 150 Aerobat

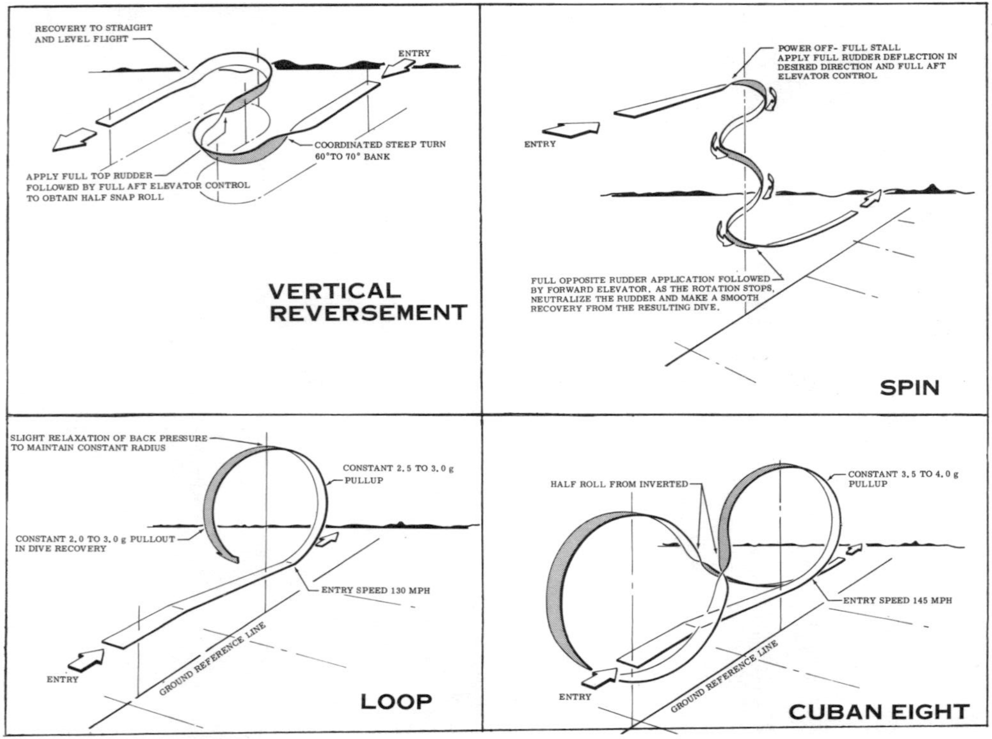
RECOVERY TO STRAIGHT AND LEVEL FLIGHT
ENTRY
COORDINATED STEEP TURN 60°TO 70° BANK
APPLY FULL TOP RUDDER FOLLOWED BY FULL AFT ELEVATOR CONTROL TO OBTAIN HALF SNAP ROLL
VERTICAL REVERSEMENT
POWER OFF- FULL STALL APPLY FULL RUDDER DEFLECTION IN DESIRED DIRECTION AND FULL AFT ELEVATOR CONTROL
ENTRY
FULL OPPOSITE RUDDER APPLICATION FOLLOWED BY FORWARD ELEVATOR. AS THE ROTATION STOPS, NEUTRALIZE THE RUDDER AND MAKE A SMOOTH RECOVERY FROM THE RESULTING DIVE.
SPIN
SLIGHT RELAXATION OF BACK PRESSURE TO MAINTAIN CONSTANT RADIUS
CONSTANT 2.5 TO 3.0 g PULLUP
CONSTANT 2.0 TO 3.0 g PULLOUT IN DIVE RECOVERY
ENTRY SPEED 130 MPH
GROUND REFERENCE LINE
ENTRY
LOOP
HALF ROLL FROM INVERTED
CONSTANT 3.5 TO 4.0 g PULLUP
ENTRY SPEED 145 MPH
GROUND REFERENCE LINE
ENTRY
CUBAN EIGHT

Miss Joyce Case, former U.S. Women's Aerobatic Champion, is special factory representative for the Cessna 150 Aerobat.

harness, tinted ceiling skylights, and a G-meter – plus the distinctive paint job.

Miss Joyce Case, champion female aerobatic pilot, is special 150 Aerobat representative for Cessna as this is written, and if you can manage a demonstration ride with Joyce, you're in luck. She's not only a very good akro-pilot (oughta be; she was handling the stick on her dad's Great Lakes before her feet could reach the rudder pedals), but she is also one of the nicest, most unaffected pretty girls you'll ever meet.

1971 Cessna. 150 & 150 AEROBAT

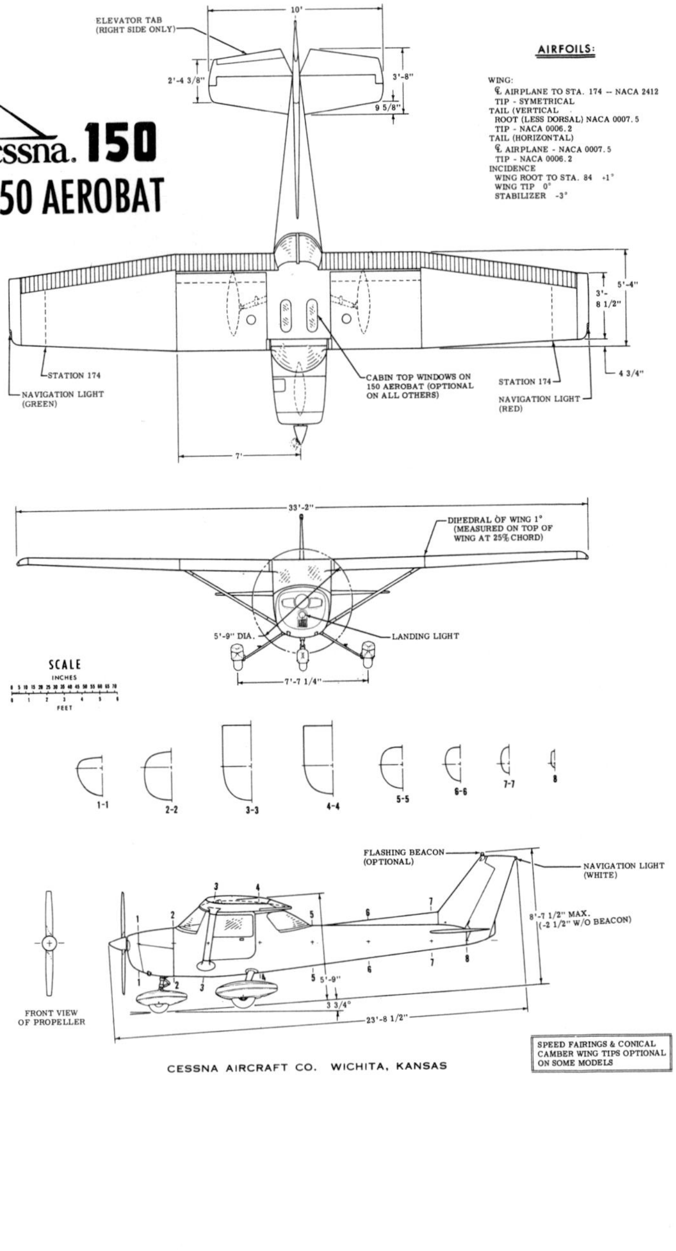

We weren't *that* lucky, but we were fortunate enough to be in Wichita the day Harold Krier (three-time Men's National Champion Aerobatic Pilot) evaluated the prototype 150 Aerobat back in June, 1969. We had dinner with Harold that evening (well, actually, hamburgers and malts at a drive-in), and he was enthusiastic over the new Aerobat.

Harold told us that, in his opinion, this craft represented an excellent blending of desirable traits in a sport/trainer: it was capable of all the fancy didos the average private pilot has any business performing; yet it retained all the forgiving and gentle qualities of its non-checkered sisters. In fact, it *was* the same as other 150's, with just a little beefing-up here and there.

Harold said he had performed snap rolls; snaps from wing-overs, slow rolls, 4-point rolls (without power), Immelmanns, loops, loops with snaps, Cuban-8's, spins and a number of inverted maneuvers that the airplane would not be certified for. He said the rudder was inadequate for some maneuvers, but explained that that is true of almost any production craft not designed exclusively for aerobatics.

Harold did surprise us when he said he hoped Cessna kept the 100-hp engine in the Aerobat; we had always had the impression that the tumble types wanted power to spare. But Krier said, "I think the 150 is a very nice little package as it stands, and I don't think Cessna should tamper with it. Give it a bigger engine and you have to modify the airframe. It gets bigger and heavier and more expensive and you no longer have a 150, but something else entirely."

Harold mentioned in passing that the 150 has the same airfoil section – NACA 2412 – as his specially-built KrierKraft (now being flown by akro-champ Charlie Hillard). He predicted that the 150 Aerobat would quickly dominate the sport/trainer/aerobatic market, and he was right about that, too.

1971 Model 150 Performance and Specifications*

GROSS WEIGHT	1600 lbs., 726 kg
SPEED	
Top Speed at Sea Level	122 mph, 196 km/h
Cruise, 75% Power @ 7000 Ft.	117 mph, 188 km/h
RANGE:	
Cruise, 75% Power @ 7000 Ft. 22.5 Gallons, No Reserve	475 mi, 764 km 4.1 hrs. 117 mph, 188 km/h
Cruise, 75% Power @ 7000 Ft. 35 Gallons, No Reserve	725 mi, 1165 km 6.2 hrs. 117 mph, 188 km/h
Optimum Range @ 10,000 Ft. 22.5 Gallons, No Reserve	565 mi, 909 km 6.1 hrs. 93 mph, 149 km/h
Optimum Range @ 10,000 Ft. 35 Gallons, No Reserve	880 mi, 1416 km 9.4 hrs. 93 mph, 149 km/h
RATE OF CLIMB AT SEA LEVEL	670 fpm, 204 m/m
SERVICE CEILING	12,650 ft, 3856 m
TAKEOFF:	
Ground Run	735 ft. 224 m
Total Distance Over 50-Ft. Obstacle	1385 ft. 422 m
LANDING:	
Ground Roll	445 ft. 136 m
Total Distance Over 50-Ft. Obstacle	1075 ft. 328 m
STALL SPEED:	
Flaps Up, Power Off	55 mph, 89 km/h
Flaps Down, Power Off	48 mph, 77 km/h
BAGGAGE	120 lbs. 54 kg
POWER LOADING	16.0 lb/hp, 7.3 kg/hp
FUEL CAPACITY: Total	
Standard Tanks	26 gal, 98 liters
Optional Long Range Tanks	38 gal, 144 liters
OIL CAPACITY	6 qts, 5.7 liters
ENGINE	100 rated hp at 2750 rpm
PROPELLER: Fixed-Pitch, Diameter	69 inches, 1.75 m
HEIGHT (With Depressed Nose Strut)	8 ft. 7 1/2 in., 2.63 m

	STANDARD	TRAINER	COMMUTER
EMPTY WEIGHT	975 lbs. 442 kg	1005 lbs. 456 kg	1060 lbs. 481 kg
USEFUL LOAD	625 lbs. 284 kg	595 lbs. 270 kg	540 lbs. 245 kg
WING LOADING	10.2 lb/ft^2 49.8 kg/m^2	10.2 lb/ft^2 49.8 kg/m^2	10.0 lb/ft^2 48.9 kg/m^2
WING SPAN	32 ft, 8 1/2 in 9.97 m	32 ft, 8 1/2 in 9.97 m	33 ft, 1 1/2 in 10.10 m
WING AREA	157 ft.^{2} 14.6 m^2	157 ft.^{2} 14.6 m^2	159.5 ft.^{2} 14.8 m^2
LENGTH	23 ft. 3 in. 7.09m	23 ft. 9 in. 7.24m	23 ft. 9 in. 7.24m

* Subject to change without notice

1971 Model 150 Aerobat Performance and Specifications*

GROSS WEIGHT	1600 lbs.	726 kg
SPEED:		
Top Speed at Sea Level	120 mph	193 kph
Cruise, 75% Power @ 7000 Ft.	115 mph	185 kph
RANGE:		
Cruise, 75% Power @ 7000 Ft.	470 mi.	756 km
22.5 Gallons, No Reserve	4.1 hrs.	4.1 hrs.
	115 mph	185 kph
Cruise, 75% Power @ 7000 Ft.	715 mi.	1151 km
35 Gallons, No Reserve	6.2 hrs.	6.2 hrs.
	115 mph	185 kph
Optimum Range @ 10,000 Ft.	555 mi.	893 km
22.5 Gallons, No Reserve	6.1 hrs.	6.1 hrs.
	91 mph	146 kph
Optimum Range @ 10,000 Ft.	855 mi.	1376 km
35 Gallons, No Reserve	9.4 hrs.	9.4 hrs.
	91 mph	146 kph
RATE OF CLIMB AT SEA LEVEL	670 fpm	204 mpm
SERVICE CEILING	12,650 ft.	3856 m
TAKEOFF:		
Ground Run	735 ft.	224 m
Total Distance Over 50-Ft. Obstacle	1385 ft.	422 m
LANDING:		
Ground Roll	445 ft.	136 m
Total Distance Over 50-Ft. Obstacle	1075 ft.	328 m
STALL SPEED:		
Flaps Up, Power Off	55 mph	89 kph
Flaps Down, Power Off	48 mph	77 kph
BAGGAGE	120 lbs.	54 kg
POWER LOADING	16.0 lb/hp	7.3 kg/hp
FUEL CAPACITY: Total		
Standard Tanks	26 gal.	98 liters
Optional Long Range Tanks	38 gal.	144 liters
OIL CAPACITY	6 qts.	5.7 liters
ENGINE	100 rated hp at 2750 rpm	
PROPELLER: Fixed-Pitch, Diameter	69 inches	1.75 m
HEIGHT (With Depressed Nose Strut)	8 ft. 7 1/2 in.	2.63 m
EMPTY WEIGHT	1030 lbs.	467 kg
USEFUL LOAD	570 lbs.	259 kg
WING LOADING	10.2 lb/ft.^{2}	49.8 kg/m^2
WING SPAN	32 ft., 8 1/2 in.	9.97 m
WING AREA	157 ft.^{2}	14.6 m^2
LENGTH	23 ft. 9 in.	7.24 m

* Subject to change without notice

Guide for Estimating Cost of Operation—150 & 150 Aerobat

	*PER YEAR	PER YEAR	*PER HOUR	PER HOUR	PER HOUR	PER HOUR
FLIGHT HOURS PER YEAR.				300	500	700
FLIGHT MILES PER YEAR: 117 mph, 75% power at 7,000 ft., Lean Mixture.				35,100	58,500	81,900
GASOLINE: 5.5 gal./hr. at 43¢/gal.				$ 2.37	$ 2.37	$ 2.37
OIL: 3/4 pt./hr. plus oil change every 25 hrs. (6 qts. per change) at 63¢/qt.				$ .39	$ .39	$ .39
AIRPLANE & ENGINE MAINTENANCE.				$ 1.25	$ 1.25	$ 1.25
TOTAL OPERATIONAL COSTS (listed above).				$ 4.01	$ 4.01	$ 4.01
RESERVE FOR ENGINE OVERHAUL.				$ .80	$ .80	$ .80
STORAGE: at $40.00 per month.		$ 480		$ 1.60	$ .91	$ .68
INSURANCE (a guide for the following coverage): All risks hull coverage (Deductibles: In motion $250.00) (Not in motion $50.00).		**$ 805				
Liability Coverage (Single limit $250,000) (Including passengers).		**$ 140				
TOTAL INSURANCE (Industrial private business & pleasure use by a qualified pilot).		**$ 945		$ 3.15	$ 1.89	$ 1.35
COST PER HOUR.				$ 9.56	$ 7.61	$ 6.84
COST PER MILE: 75% power, 117 mph.				$ .082	$ .065	$ .058
COST PER SEAT MILE: 2 People.				$ .041	$ .033	$ .029

FOR DEPRECIATION: Consult your tax authority. To estimate the per hour or per mile cost, divide annual depreciation amount by annual hours or miles.

*YOUR COSTS: No doubt vary from these national averages. This column is included for you to compute your estimate of your hours, miles and costs.

**INSURANCE COSTS FOR AEROBAT: Available on individual application only.

CHAPTER 8: CESSNA 182/SKYLANE, 205 AND SUPER SKYLANE

The Cessna 182 first appeared in 1956. It was a Model 180 with tricycle landing gear. Except for "styling" differences since that time, the same could be said for the 182's and Skylanes (the Skylane is the posh version of the 182) built through 1971. The 182/Skylane represents a substantial step upward from the 172/Skyhawk series, and the transition is not lightly made, because the Skylane does not fly or land like a Skyhawk. It's five hundred pounds more airplane, with more power (efficiently applied via a constant-speed prop), and it insists that its new owner stops grinning and starts thinking. It's not a student airplane; it's for serious cross-country transportation – any kind of country.

Since we didn't have a 1971 Model Skylane handy, we took a demonstration ride in a '69 Model, because there was little change in this craft during these three years, except for a slight weight increase (we were surprised that the '71 did not switch to tubular-steel gear legs). However, even that didn't turn out very well because we were up-tight for time and had only a couple of hours to go on a raw November day. That was decidedly unfair to an airplane that likes to get up high and go far. Our veteran Cessna man at Lawton (Oklahoma) Municipal, Ray Johnson, says that no one can really appreciate the Skylane until he's taken it on a long cross-country flight, preferably, to California or Oregon and back with some stops at high-altitude airports in the West.

As we said, the first (1956) 182 was a 180 with tricycle landing gear. In 1957, the gear legs were shortened, and a year later the Skylane version appeared. The swept tail was added in 1960; in '61 came an additional window on each side, and in 1962 the big rear window for 360-degree vision, plus electric flaps and a wider cabin. A lot of interior improvements followed during the next five years, then the 1969 Model again had its main gear legs shortened and the space between the wheels expanded. But since the nose-wheel could not be lowered (due to the need for prop clearance with the ground), this resulted in a slightly nose-high stance for the 182/Skylane, a situation that makes for easier entry and exit from the cabin, perhaps better rough-field handling and a lower profile to better handle cross-wind landings. But it doesn't do anything for the pilot transitioning from the 172, because, if he drives the 182 down to land like some people do the Super Car, he's liable to bend the firewall a little. The 182/Skylane must be brought in for a nose-high touch-down on the main wheels.

Of course, the 182 pilots will be quick to point out that that's the way we *should* have been landing the 172 and 150 (experienced flyers are like reformed sinners; they have small tolerance for those of us who are doing things the way *they* used to).

Anyway, the point is, the 182/Skylane is not going to cover up for you or shrug off indifferent control handling. She steps higher than the 172 and requires a tighter rein.

This is not to say that the Skylane is hard to fly or tricky. It isn't. It's still a Cessna. But the controls are heavier and you know you're flying more airplane. The principal differences for the pilot stepping up from the 150 or 172 will probably be in landings and power management. That "extra throttle" – prop control – gives fresh significance to the tachometer and adds a manifold pressure gauge.

1971 Cessna Skylane

Engine starting procedure is much like that of the 172 except that you have cowl flaps to open and prop pitch control to push in for "High RPMs." The pre-take-off run-up is done at the usual 1,700 rpms (with prop in high rpms), and in addition, you'll "exercise" the prop a couple of times – pull out the knob changing pitch from high to low rpms – to insure oil circulation that activates pitch change. Then, leaving the prop in high rpms, feed-in full throttle for take-off.

You'll notice the extra power, and 60 mph will come quickly. You'll also notice that it requires a firmer pressure to lift the nose-wheel off if you've been flying a Skyhawk.

Maximum power for the take-off climb should be limited to that absolutely necessary for safety, according to the Skylane Owner's Manual, so we reduce power to the recommended 23 inches of manifold pressure and 2,450 rpms (about 75% power) for climb-out at 100 mph. This gives us 1,300 fpm with the OAT registering 44 degrees F. We are almost 50 lbs light, however. The book says we can get 1,700 fpm at this loading with full power and an IAS of 85 mph.

The Skylane gives quite a bit of flexibility in cruise and range at various altitudes. Fully loaded, at 2,500 feet, turning 2,450 with 23 inches, TAS is 156 mph and range (with 60 gal) is 660 miles. This can be stretched to 865 miles at this inefficient altitude if you cut rpms to 2,000 and manifold pressure to 17 inches with a TAS of 100 mph. But you aren't likely to ever go anywhere at that altitude in a Skylane. Seventy-five hundred feet is much better; 10,000 is probably best, and you can go to 15,000 if you have oxygen. At 15,000, 2,450 rpms and 16 inches will give you 145 mph TAS and a range of 835 miles.

For a given throttle setting, select the lowest rpms (with the prop control) in the green arc range that will give smooth engine operation. Cowl flaps should be adjusted to maintain cylinder head temperature at approximately 2/3 of the normal operating range (green arc); and to lean the mixture for the best range, pull out the mixture control until the engine becomes rough, then push it back in just far enough to smooth out the engine again. Any change in altitude, power setting or carburetor heat will require a change in the lean mixture setting.

It's possible to land the Skylane without use of trim, but that's doing it the hard way. And although the book says to bring it in power-off with 40-degrees of flap at 69 mph for short-field landings, this produces a fairly high sink rate and you've got to be right on top of it to make your flare just right or you'll touch-down pretty hard. Seems to us the best way is to hold a little power and flatten-out the final, gradually backing-off the throttle and closing it just a little before flare. Then hold the wheel well back to insure that you touch-down on the main gear. It's best to hold the nose-wheel off for the first couple of hundred feet of the roll-out, unless you have a cross-wind to contend with.

The 1970 Model 182/Skylane adopted the conical cambered wingtips and new elevator control system to lower elevator stick forces. Basic list price of the 182 was $19,795. The Skylane was priced at $20,895. Used prices range all the way down to about $6,000 for an early Model 182.

The Cessna 205 was produced 1962-1964 inclusive. It has been described as a stretched Skylane with an additional 30-hp and fuel injection. It was also called, at the time it appeared, a Cessna 210 with fixed landing gear. Anyway, less than 600 were built during those three years before it was replaced by – or evolved into – the Super Skylane.

Prices, principal specs and performance figures for the 205 are listed in the Model Comparison Tables following Chapter 10.

The Super Skylane and Turbo Super Skylane would be easier for Cessna watchers to fit into their proper categories if Cessna would only call them "Super 206's" instead of Skylanes because, it seems to us, these craft aren't really Skylanes, but a definite step upward – in power, size and weight – to another class. They are

in fact up-town versions of the 206 Skywagon. For the extra conveniences, snootier name and different paint scheme, you pay $1,400 ($25,995 for the '70 Skywagon 206 vs. $27,450 for the '70 Super Skylane), and this would seem to lend weight to our long-standing suspicion that Cessna dealers need only to ask, "How much money have you got?" in order to roll out an airplane and match that figure almost to the dollar. (We can just picture the scene: "But I've only got $27,406.09," the customer says. "That's all right, sir," the dealer replies and, picking up the phone asks for his service department. "Okay, Mac, take a seat out of that new Super Skylane. A gentleman here wants a five-placer.")

Seriously, Cessna *does* effectively cover the gen-av market in gradual steps, from the 150 to the Citation 500 Fan Jet. This encompasses a price range from $8,895 to $695,000. So if a Cessna

1970 Cessna Super Skylane

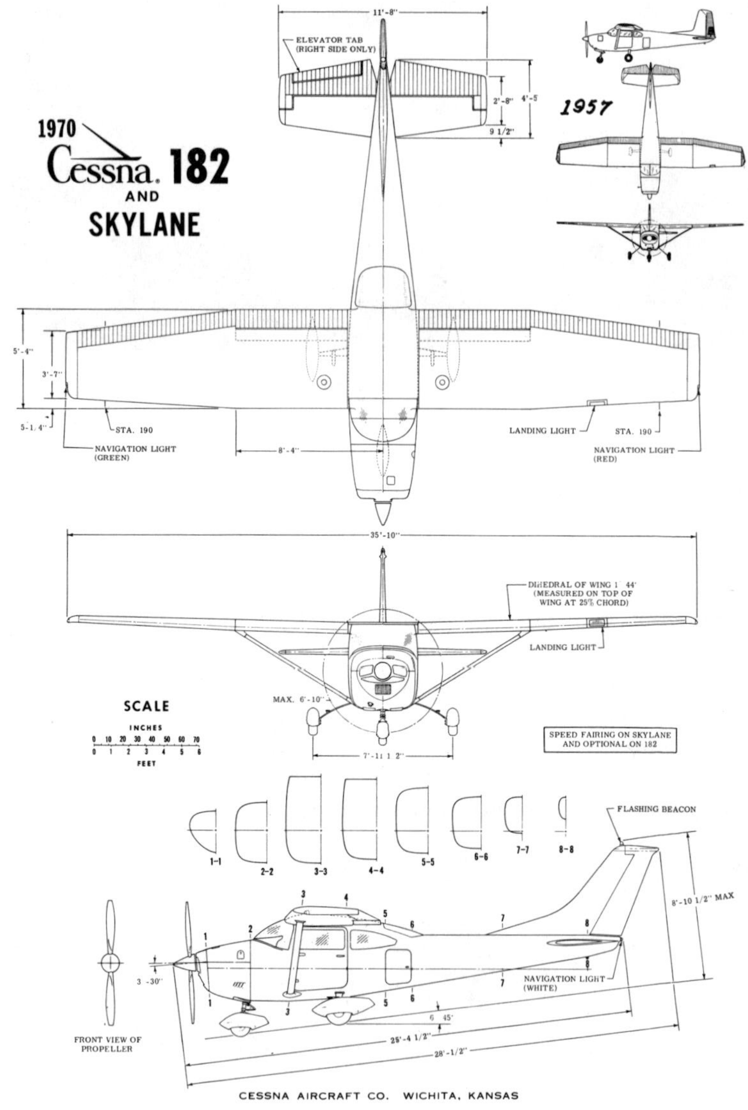

1970
Cessna. 182
AND
SKYLANE
1957
11'-8"
ELEVATOR TAB
(RIGHT SIDE ONLY)
2'-8"
4'-5"
9 1/2"
5'-4"
3'-7"
5-1/4"
STA. 190
NAVIGATION LIGHT
(GREEN)
8'-4"
LANDING LIGHT
STA. 190
NAVIGATION LIGHT
(RED)
35'-10"
DIHEDRAL OF WING 1° 44'
(MEASURED ON TOP OF
WING AT 25% CHORD)
LANDING LIGHT
MAX. 6'-10"
7'-11 1/2"
SPEED FAIRING ON SKYLANE
AND OPTIONAL ON 182
SCALE
INCHES
0 10 20 30 40 50 60 70
0 1 2 3 4 5 6
FEET
1-1
2-2
3-3
4-4
5-5
6-6
7-7
8-8
FLASHING BEACON
8'-10 1/2" MAX
3°-30"
NAVIGATION LIGHT
(WHITE)
6° 45'
25'-4 1/2"
28'-1/2"
FRONT VIEW OF
PROPELLER
CESSNA AIRCRAFT CO. WICHITA, KANSAS

dealer hasn't got what you want, perhaps there's nothing left to do but join the EAA and build your own.

The Super Skylane was introduced late in 1964 (1965 Model), and was intended for the charter and air taxi operators. It would comfortably seat six adults (with plenty of leg room) and then take-off with full tanks and nearly 200 lbs of luggage. At optimum altitude, it would haul this load nearly 600 miles (with reserve fuel) at 160 mph or better, and it would do so economically, its fuel-injected 285-hp Continental 10-520-A engine consuming about 15 gallons of gas per hour.

By 1970, the Super Skylane had gained 300 lbs at gross weight and had an increased useful load of 250 lbs – at the cost of a slight amount of speed/range/initial climb. It seemed an excellent airplane for its task, but its future appeared uncertain because the Cessna 207 – for very little more money – could do even more with the same engine.

The Turbo Super Skylane, introduced in 1966, is equipped with a turbo supercharger. It has less useful load than its unboosted sister, but will get up high and go fast.

The supercharger helps the engine breathe more efficiently when the temperature is hot and the density altitude is high. It provides 32.5 inches of manifold pressure up to 19,000 ft. Above that, it keeps working like an extra set of lungs compressing the thin air into a richer source of oxygen for the engine. As a result, the Turbo Super Skylane can effectively operate off the highest airfields in the world fully loaded.

This airplane will climb out at 1,030 fpm initially, and has a minimum loss of horsepower as altitude is gained. The turbo system continues to deliver cruise power up to 24,000 ft. There are no continuous control or throttle adjustments during climb, and no additional gauges to watch. The fuel-injected, 285-hp Continental engine (TS10-520-C) is specially adapted for turbocharging, and the turbine is automatically regulated by an absolute pressure controller and overboost control valve.

A two-to-three hour oxygen system is standard equipment on the Turbo Super Skylane. This installation includes an overhead console with capacity gauge and system control, six individual ports, control wheel microphone switch, external filler valve, and a 76-cubic-inch oxygen bottle.

Suggested list price of the Super Skylane for 1970 was 27,450. The Turbo Super Skylane for 1970 was priced at $31,950.*

*We do not attempt to discuss the various avionics installations available for Cessna airplanes due to space limitations in this book. For a complete rundown on light aircraft avionics and their uses, see the Modern Aircraft Series book, "NAV/COM GUIDE FOR PILOTS," by Don Downie.

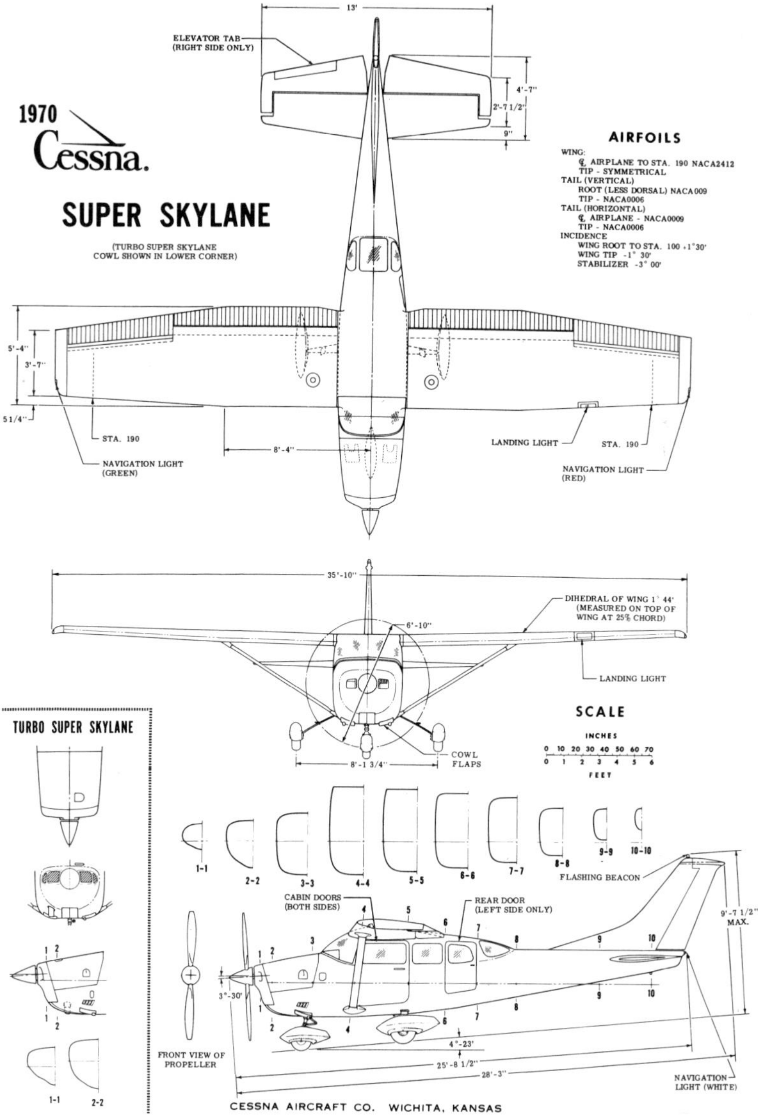

1970
Cessna.
SUPER SKYLANE
(TURBO SUPER SKYLANE COWL SHOWN IN LOWER CORNER)
13'
ELEVATOR TAB (RIGHT SIDE ONLY)
4'-7"
2'-7 1/2"
9"
AIRFOILS
WING:
℄ AIRPLANE TO STA. 190 NACA2412
TIP - SYMMETRICAL
TAIL (VERTICAL)
ROOT (LESS DORSAL) NACA009
TIP - NACA0006
TAIL (HORIZONTAL)
℄ AIRPLANE - NACA0009
TIP - NACA0006
INCIDENCE
WING ROOT TO STA. 100 +1°30'
WING TIP -1° 30'
STABILIZER -3° 00'
5'-4"
3'-7"
5 1/4"
STA. 190
8'-4"
NAVIGATION LIGHT (GREEN)
LANDING LIGHT
STA. 190
NAVIGATION LIGHT (RED)
35'-10"
6'-10"
DIHEDRAL OF WING 1° 44' (MEASURED ON TOP OF WING AT 25% CHORD)
LANDING LIGHT
8'-1 3/4"
COWL FLAPS
SCALE
INCHES
0 10 20 30 40 50 60 70
0 1 2 3 4 5 6
FEET
TURBO SUPER SKYLANE
1-1
2-2
1-1
2-2
3-3
4-4
5-5
6-6
7-7
8-8
9-9
10-10
FLASHING BEACON
CABIN DOORS (BOTH SIDES)
REAR DOOR (LEFT SIDE ONLY)
9'-7 1/2" MAX.
3°-30'
FRONT VIEW OF PROPELLER
4°-23'
25'-8 1/2"
28'-3"
NAVIGATION LIGHT (WHITE)
CESSNA AIRCRAFT CO. WICHITA, KANSAS

1970 Skylane and Model 182 Performance and Specifications*

GROSS WEIGHT	**2950 lbs.	1338 kg
SPEED:		
Top speed at Sea Level	168 mph	270 kph
Cruise, 75% Power at 6500 ft.	160 mph	257 kph
RANGE:		
Cruise, 75% Power at 6500 ft.	690 mi.	1110 km
60 Gallons, No Reserve	4.3 hr.	4.3 hr.
	160 mph	257 kph
Cruise, 75% Power at 6500 ft.	910 mi.	1464 km
79 Gallons, No Reserve	5.7 hr.	5.7 hr.
	160 mph	257 kph
Optimum Range at 10,000 ft.	1160 mi.	1867 km
79 Gallons, No Reserve	10.1 hr.	10.1 hr.
	115 mph	185 kph
RATE-OF-CLIMB AT SEA LEVEL	890 fpm	271 mpm
SERVICE CEILING	17,700 ft.	5395 m
TAKEOFF:		
Ground Run	705 ft.	215 m
Total Distance Over 50-ft. Obstacle	1350 ft.	411 m
LANDING:		
Ground Roll	590 ft.	180 m
Total Distance Over 50-ft. Obstacle	1350 ft.	411 m
STALL SPEED:		
Flaps Up, Power Off	66 mph	106 kph
Flaps Down, Power Off	57 mph	92 kph
EMPTY WEIGHT: (Approx.)		
Skylane	1640 lbs.	744 kg
Standard Model 182	1580 lbs.	717 kg
USEFUL LOAD:		
Skylane	1310 lbs.	594 kg
Standard Model 182	1370 lbs.	621 kg
BAGGAGE	120 lbs.	54 kg
WING LOADING	16.9 lbs/ft^2	82.6 kg/m^2
POWER LOADING	12.8 lbs/hp	5.8 kg/hp
FUEL CAPACITY: Total		
Standard Tanks	65 gal.	246 liters
Optional Long-Range Tanks	84 gal.	318 liters
OIL CAPACITY	12 quarts	11.4 liters
POWER:		
Six-Cylinder Engine	230 rated hp at 2600 rpm	
PROPELLER:		
Constant-Speed, Diameter	82 in.	2.08 m
WING SPAN	35 ft. 10 in.	10.92 m
WING AREA	174 sq. ft.	16.2 m^2
LENGTH	28 ft. 1/2 in.	8.55 m
HEIGHT (With Depressed Nose Strut)	8 ft. 10-1/2 in.	2.71 m

*Performance data is shown for the Skylane which is 2 to 3 mph faster than a standard-equipped Model 182 (without speed fairings). While all other performance figures are the same for the 182 as shown for the Skylane, there is a corresponding difference in range.

**Maximum takeoff weight is 2950 lbs., maximum landing weight is 2800 lbs. (1270 kg).

1970 Super Skylane Performance and Specifications

GROSS WEIGHT	3600 lbs., 1633 kg
SPEED: Best Power Mixture	
Top Speed at Sea Level	174 mph, 280 kph
Cruise, 75% Power at 6000 ft.	163 mph, 262 kph
RANGE: Normal Lean Mixture	
Cruise, 75% Power at 6000 ft. 63 Gallons, No Reserve	650 miles, 1045 km 4.0 hours 162 mph, 261 kph
Cruise, 75% Power at 6000 ft. 80 Usable Gallons, No Reserve	830 miles, 1335 km 5.1 hours 162 mph, 261 kph
Optimum Range at 10,000 ft. 80 Usable Gallons, No Reserve	1020 miles, 1640 km 7.6 hours 134 mph, 216 kph
RATE-OF-CLIMB AT SEA LEVEL	920 fpm, 280 mpm
SERVICE CEILING	14,800 ft., 4511 m
TAKEOFF:	
Ground Run	910 ft., 277 m
Total Distance Over 50-ft. Obstacle	1810 ft., 552 m
LANDING:	
Ground Roll	735 ft., 224 m
Total Distance Over 50-ft. Obstacle	1395 ft., 425 m
STALL SPEED:	
Flaps Up, Power Off	70 mph, 113 kph
Flaps Down, Power Off	61 mph, 98 kph
EMPTY WEIGHT: (Approx.)	1835 lbs., 832 kg
USEFUL LOAD	1765 lbs., 801 kg
WING LOADING	20.7 lbs./ft.^{2}, 100.8 kg/m^2
POWER LOADING	12.6 lbs./hp, 5.7 kg/hp
FUEL CAPACITY: Total	
Standard Tanks	65 gallons, 246 liters
Optional Long-Range Tanks	84 gallons, 318 liters
OIL CAPACITY: Total	12 quarts, 11.4 liters
POWER	Six-Cylinder, Fuel-Injection Engine, 285 Rated HP at 2700 RPM
PROPELLER:	
*Two-Bladed, Constant-Speed - Diameter	82 inches, 2.08 m
WING AREA	174 ft.^{2}, 16.2 m^2
LENGTH	28 ft. 3 in., 8.61 m
HEIGHT: (With Depressed Nose Strut)	9 ft. 7-1/2 in., 2.93 m

* Performance for optional three-bladed propeller is essentially the same as above.

Guide for Estimating Cost of Operation—Super Skylane

	•PER YEAR	PER YEAR	•PER HOUR	PER HOUR	PER HOUR	PER HOUR
FLIGHT HOURS PER YEAR.				300	500	700
FLIGHT MILES PER YEAR: 162 mph, 75% power at 6,000 ft., Normal Lean Mixture.				48,600	81,000	113,400
GASOLINE: 15.8 gal./hr. at 46¢/gal.				$ 7.27	$ 7.27	$ 7.27
OIL: 1 pt./hr. plus oil change every 25 hrs. (12 qts. per change) at 63¢/qt.				$ .63	$ .63	$ .63
AIRPLANE & ENGINE MAINTENANCE.				$ 2.13	$ 2.13	$ 2.13
TOTAL OPERATIONAL COSTS (listed above).				$10.05	$10.05	$10.05
RESERVE FOR ENGINE OVERHAUL.				$ 1.98	$ 1.98	$ 1.98
STORAGE: at $50.00 per month.		$ 600		$ 2.00	$ 1.20	$ .86
INSURANCE (a guide for the following coverage): All risks hull coverage (Deductibles: In motion $250.00) (Not in motion $50.00).		$1,582				
Liability Coverage (Single limit $250,000) (Including passengers).		$ 140				
TOTAL INSURANCE (Industrial private business & pleasure use by a qualified pilot).		$1,722		$ 5.74	$ 3.44	$ 2.46
COST PER HOUR.				$19.92	$16.82	$15.50
COST PER MILE: 75% power, 162 mph.				$ .123	$ .104	$ .096
COST PER SEAT MILE: 6 People.				$ .021	$ .017	$ .016

FOR DEPRECIATION: Consult your tax authority. To estimate the per hour or per mile cost, divide annual depreciation amount by annual hours or miles.

•YOUR COSTS: No doubt vary from these national averages. This column is included for you to compute your estimate of your hours, miles and costs.

1970 Turbo Super Skylane Performance and Specifications

GROSS WEIGHT	3600 lbs., 1633 kg
SPEED: Best Power Mixture	
Top Speed at 19,000 ft.	200 mph, 322 kph
Cruise, 75% Power at 24,000 ft.	184 mph, 296 kph
Cruise, 75% Power at 10,000 ft.	170 mph, 274 kph
RANGE: Normal Lean Mixture	
Cruise, 75% Power at 24,000 ft. 63 Usable Gallons, No Reserve	700 miles, 1127 km 3.8 hours 182 mph, 293 kph
Cruise, 75% Power at 10,000 ft. 63 Usable Gallons, No Reserve	645 miles, 1038 km 3.8 hours 168 mph, 270 kph
Cruise, 75% Power at 24,000 ft. 80 Usable Gallons, No Reserve	890 miles, 1432 km 4.9 hours 182 mph, 293 kph
Cruise, 75% Power at 10,000 ft. 80 Usable Gallons, No Reserve	820 miles, 1320 km 4.9 hours 168 mph, 270 kph
Optimum Range at 15,000 ft. 80 Usable Gallons, No Reserve	1050 miles, 1690 km 7.6 hours 139 mph, 224 kph
RATE-OF-CLIMB AT SEA LEVEL	1030 fpm, 314 mpm
SERVICE CEILING	26,300 ft., 8020 m
TAKEOFF:	
Ground Run	910 ft., 277 m
Total Distance Over 50-ft. Obstacle	1810 ft., 552 m
LANDING:	
Ground Roll	735 ft., 224 m
Total Distance Over 50-ft. Obstacle	1395 ft., 425 m
STALL SPEED:	
Flaps Up, Power Off	70 mph, 113 kph
Flaps Down, Power Off	61 mph, 98 kph
EMPTY WEIGHT: (Approx.)	1935 lbs., 878 kg
USEFUL LOAD	1665 lbs., 755 kg
WING LOADING	20.7 lbs./ft.^{2}, 100.8 kg/m^2
POWER LOADING	12.6 lbs./hp, 5.7 kg/hp
FUEL CAPACITY: Total	
Standard Tanks	65 gallons, 246 liters
Optional Long-Range Tanks	84 gallons, 318 liters
OIL CAPACITY: Total	13 quarts, 12.3 liters
POWER	Six-Cylinder, Turbosuper-charged, Fuel-Injection Engine, 285 Rated HP at 2700 RPM
PROPELLER:	
*Two-Bladed, Constant-Speed - Diameter	82 inches, 2.08 m
WING SPAN	35 ft. 10 in., 10.92 m
WING AREA	174 ft.^{2}, 16.2 m^2
LENGTH	28 ft. 3. in., 8.61 m
HEIGHT: (With Depressed Nose Strut)	9 ft. 7-1/2 in., 2.93 m

*Performance for optional three-bladed propeller is essentially the same as above.

Guide for Estimating Cost of Operation—Turbo Super Skylane

	*PER YEAR	PER YEAR	*PER HOUR	PER HOUR	PER HOUR	PER HOUR
FLIGHT HOURS PER YEAR.				300	500	700
FLIGHT MILES PER YEAR: 168 mph, 75% power at 10,000 ft., Normal Lean Mixture. .				50,400	84,000	117,600
GASOLINE: 16.6 gal./hr. at 46¢/gal.				$ 7.64	$ 7.64	$ 7.64
OIL: 1 pt./hr. plus oil change every 50 hrs. (13 qts. per change) at 63¢/qt.				$ .49	$ .49	$ .49
AIRPLANE & ENGINE MAINTENANCE.				$ 2.40	$ 2.40	$ 2.40
TOTAL OPERATIONAL COSTS (listed above).				$10.53	$10.530	$10.53
RESERVE FOR ENGINE OVERHAUL.				$ 2.57	$ 2.57	$ 2.57
STORAGE: at $50.00 per month.		$ 600		$ 2.00	$ 1.20	$ .86
INSURANCE (a guide for the following coverage): All risks hull coverage (Deductibles: In motion $250.00) (Not in motion $50.00).		$1,808				
Liability Coverage (Single limit $250,000) (Including passengers).		$ 140				
TOTAL INSURANCE (Industrial private business & pleasure use by a qualified pilot).		$1,948		$ 6.49	$ 3.90	$ 2.78
COST PER HOUR.				$21.59	$18.20	$16.74
COST PER MILE: 75% power, 168 mph.				$ .129	$ .108	$ .100
COST PER SEAT MILE: 6 People.				$ .022	$ .018	$ .017

FOR DEPRECIATION: Consult your tax authority. To estimate the per hour or per mile cost, divide annual depreciation amount by annual hours or miles.

*YOUR COSTS: No doubt vary from these national averages. This column is included for you to compute your estimate of your hours, miles and costs.

CHAPTER 9: THE CESSNA SKYWAGONS, MODELS 206 AND 207

The first Cessna 206 was the 1965 Model, at first called the "Super Skywagon." It was directy descended from the Model 205 which had been produced 1962-1964 inclusive. The 206 in fact was built around the same basic airframe as the 205, though the fuselage was beefed-up to permit heavy floor lads and allow a 3½-ft double cargo door. It used the Model 210D Centurion wing (introduced on the 1964 Model 210) with 18.9 feet of flap and Frise type ailerons of increased chord. Its tail group was a slightly-modified version of the 210 empennage.

In 1966 the Turbo Super Skywagon was added to Cessna's "Utiline" series making a total of four single-engine utility craft. These were the Models 180, 185 Skywagon, Super Skywagon and Turbo Super Skywagon. These latter two aircraft became, respectively, the Skywagon 206 and the Turbo Skywagon 206 in 1970 because the Model 207 was in the market by then and it would have had to be, presumably, the "Super Duper Skywagon" had the smaller craft kept the name "Super Skywagon."

If you find all this confusing (as we do), it's clearly because Cessna (as well as Cessna's competitors) has some allegedly-bright young men – with sideburns and loud neckties – who are well paid to mis-name things. Unfortunately, their often ill-invented and not-necessarily-descriptive terms also include many component parts of Cessna airplanes. We say, "unfortunately" because we think this

1970 Cessna Turbo Skywagon 206

cheapens a good product. A "Utiline" fitted with a "Quick-Scan" panel of "Space Age gray" featuring "Accru-Measure" fuel gauges, sounds like something dreamed up in Detroit rather than in Wichita. After all, the man who can write you a check for thirty thousand iron men in payment for a flying machine, presumably possesses a smattering of sophistication and a smidgen of savvy. We doubt that he's much impressed with Madison Avenue verbiage.

And we'll apologize to Cessna for voicing this pet peeve just as soon as we hear a Cessna owner refer to his "Broad Span" elevators or his "Sure-Grip" control knobs as such.

The above beef, of course, has nothing to do with the quality and performance of Cessna airplanes; and the Skywagon 206 and Turbo Skywagon 206, with the ability to operate on wheels, floats on skiis, are indeed versatile and economical utility craft possessing cabins of 120-cubic-feet capacities. This translates into a payload of six bigger-than-average people (allowing plenty of leg room and even a narrow aisle between the seats) and 300 lbs of baggage. For additional volume capacity, an optional 300-lb cargo pack may be added beneath the fuselage. All this may be transported more than 500 miles (with reserve fuel remaining), at 165 mph in the 206; and the Turbo 206 will carry just 100 lbs less approximately the same distance at speeds up to 200 mph, with an optimum cruise of 184 mph.

The 1970 Models were announced in February of that year, and improvements included a lower, re-designed cowl, light-weight

1970 Cessna Skywagon 207

conical-cambered wingtips (which the side-burned boys in the pink shirts are now calling, "Stabili-Tips"), plus fore-and-aft adjustment of the fifth and sixth seats, along with a few other minor innovations.

Factory suggested price of the Skywagon 206 for 1970 was $25,995; and the '70 Model Turbo Skywagon 206 had a basic list of $30,495.

The Cessna Skywagon 207, introduced in 1969, is the real minibus of the industry. With a 14-ft cabin, the 207 provides 156 cubic feet inside that comfortably hauls seven any-sized adults and, apparently, all their worldly goods. In addition to luggage space in the aft cabin and a 300-lb-capacity optional cargo pod slung beneath the fuselage, the 207 also has a 120-lb. capacity forward luggage compartment in its stretched nose between the firewall and cabin. And since this working airplane has an economical cruise that consumes as little as 10 gallons of fuel per hour, any air taxi operator that doesn't make money with it, just ain't tryin'.

Actually, this may be slightly over-stating the case for the 207. It has a useful load of 1,920 lbs (which is 40 lbs more than its empty dry weight), but this is without seats, avionics and fuel. Full tanks and all necessary equipment will leave about 1,400 lbs for people and baggage; but even that transposes into seven 170-pounders and 200 lbs of luggage. The loading limits of this craft are such that one may distribute this (approximate) 1,400 lbs pretty much as he pleases, although the plane does fly better with some load well aft.

1969 Cessna Skywagon 207

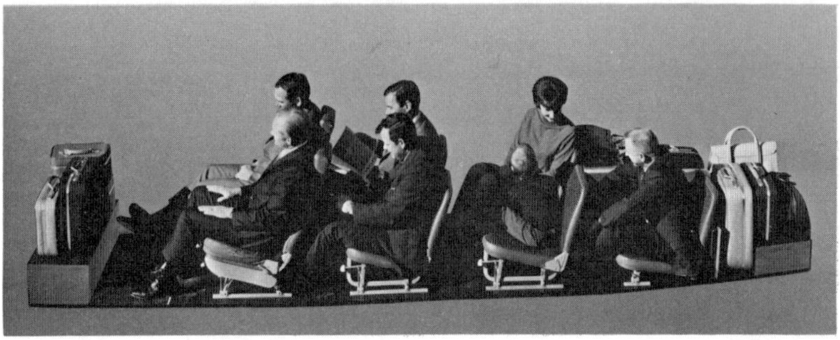

Both 207 Models have the tubular-steel gear legs and the resulting luxury ground-ride.

The Turbo Skywagon 207 can operate easily from high altitude airports and can handle non-standard days with an extra margin of safety. At its 3,800-lb gross weight, it climbs 885 fpm at sea level and has a 24,500-ft service ceiling. The Turbo 207 can reach a top speed of 189 mph and a cruise speed of 176 mph at 75% power at 20,000 feet. Range is 725 miles (with reserve fuel remaining) at 20,000 and 75%.

Our evaluation of the Turbo 207 came on a damp and overcast November day when we were joined by two newspaper reporters and three others, plus the Cessna pilot, for a purely local flight.

Since the turbo system best reveals its muscle on hot days and/or in the high places, this wasn't a very good demonstration of the difference it makes. However, we did have seven people aboard, and with a good deal of optional equipment installed and about 50 gallons of fuel, we were within 150-175 lbs of the plane's maximum allowable weight. Ground temperature was 39 degrees F, wind was 3-5 mph down the runway and airport elevation 1,108 ft.

Engine-start procedure is simple. With electrical switches off and cowl flaps open, turn on the master switch and set parking brake. Then, mixture "Full Rich," prop in "High RPMs;" with throttle closed, punch the "Lo" side of the auxiliary fuel pump switch, turn ignition key to "Start," and slowly advance the throttle. With the engine idling satisfactorily, turn off the aux fuel pump (which acts as an automatic primer when cranking the engine), turn on radios and electrics.

The auxiliary fuel pump, which comes with fuel-injected engines, plus the constant speed prop, cowl flaps and rudder trim, are the new controls to get used to if you're transitioning from the smaller or older Cessnas.

With the 13-item pre-take-off check completed, our demo pilot dropped 10 degrees of flap and applied full power. He came back on the control column as we reached 70 mph and we were airborne. Seconds later, the rate-of-climb needle settled on 900 fpm as our pilot cut back to 75% power and adjusted the climb with the wheel at a shade under 100 mph IAS. He leaned the mixture slightly as we climbed, but nothing else changed – except a slowly-increasing air speed – and we were still showing 900 fpm, at 100 mph, when we leveled off at 5,000. This of course represented a gain in TAS, with no loss in climb rate, at that altitude.

Our density altitude was a bit lower than that showing on the altimeter, and in cruise TAS worked out to 160 mph turning 2,500 rpms with 25 inches of manifold pressure, while the fuel flow meter registered 16 gallons per hour fuel consumption.

We weren't shown any stalls, because three of our passengers were new to "little" airplanes and our pilot – wisely – wanted to give them a smooth, confidence-building ride. He told us that 207 stalls are sedate, almost leisurely, and contain no surprises. Since we were impressed with his deliberate and conservative handling of the airplane, we decided his assessment would be accurate.

He showed us two landings (the second one evidently to prove that his "grease job" on the first wasn't an accident), and the way he did it, it sure looked easy. His approach was at 95 mph, slowing to 85 mph on final with full flaps. Then an unhurried, graceful flare, with throttle closed, brought touchdown on the main wheels at 65 mph. Roll-out required about 500 feet.

Suggested price of the 1970 Skywagon 207 for 1970 was $28,250 and the Turbo Skywagon 207 had a list of $32,975.

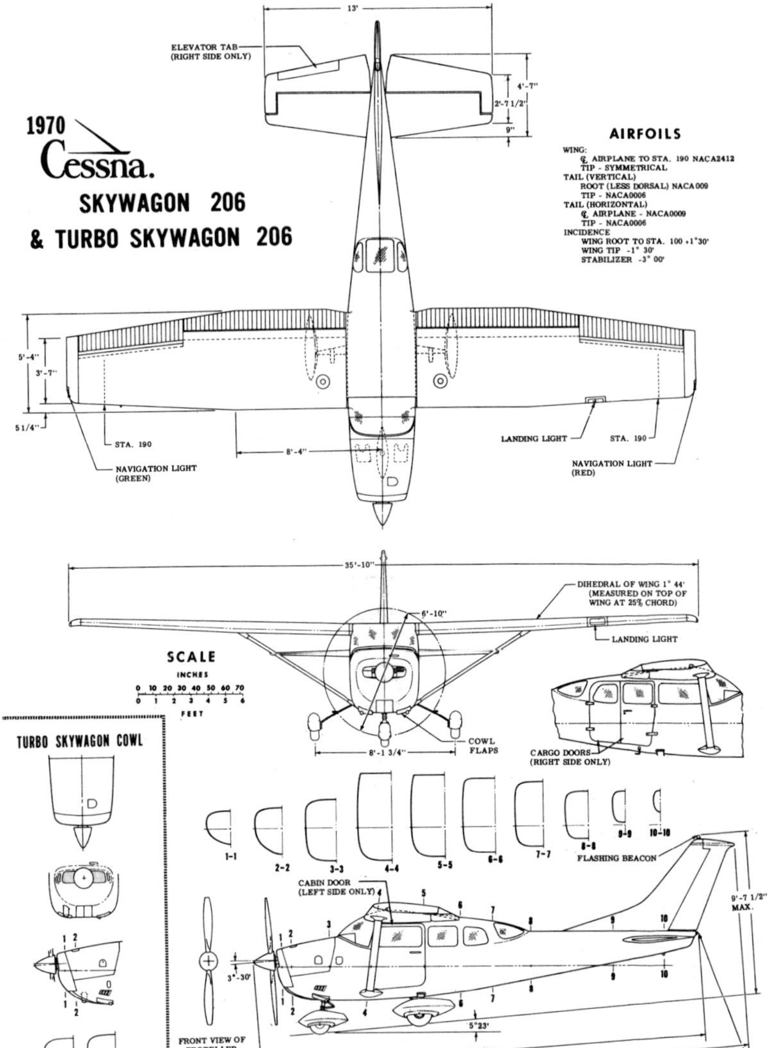
1970 Cessna.
SKYWAGON 206
& TURBO SKYWAGON 206
13'
ELEVATOR TAB (RIGHT SIDE ONLY)
4'-7"
2'-7 1/2"
9"
AIRFOILS
WING:
℄ AIRPLANE TO STA. 190 NACA2412
TIP - SYMMETRICAL
TAIL (VERTICAL)
ROOT (LESS DORSAL) NACA 009
TIP - NACA0006
TAIL (HORIZONTAL)
℄ AIRPLANE - NACA0009
TIP - NACA0006
INCIDENCE
WING ROOT TO STA. 100 +1°30'
WING TIP -1° 30'
STABILIZER -3° 00'
5'-4"
3'-7"
5 1/4"
STA. 190
NAVIGATION LIGHT (GREEN)
8'-4"
LANDING LIGHT
STA. 190
NAVIGATION LIGHT (RED)
35'-10"
DIHEDRAL OF WING 1° 44' (MEASURED ON TOP OF WING AT 25% CHORD)
6'-10"
LANDING LIGHT
SCALE
INCHES
0 10 20 30 40 50 60 70
0 1 2 3 4 5 6
FEET
TURBO SKYWAGON COWL
COWL FLAPS
8'-1 3/4"
CARGO DOORS (RIGHT SIDE ONLY)
1-1
2-2
3-3
4-4
5-5
6-6
7-7
8-8
9-9
10-10
FLASHING BEACON
CABIN DOOR (LEFT SIDE ONLY)
9'-7 1/2" MAX.
3°-30'
5°23'
FRONT VIEW OF PROPELLER
25'-8 1/4"
28'-0"
NAVIGATION LIGHT (WHITE)
1-1
2-2
CESSNA AIRCRAFT CO. WICHITA, KANSAS

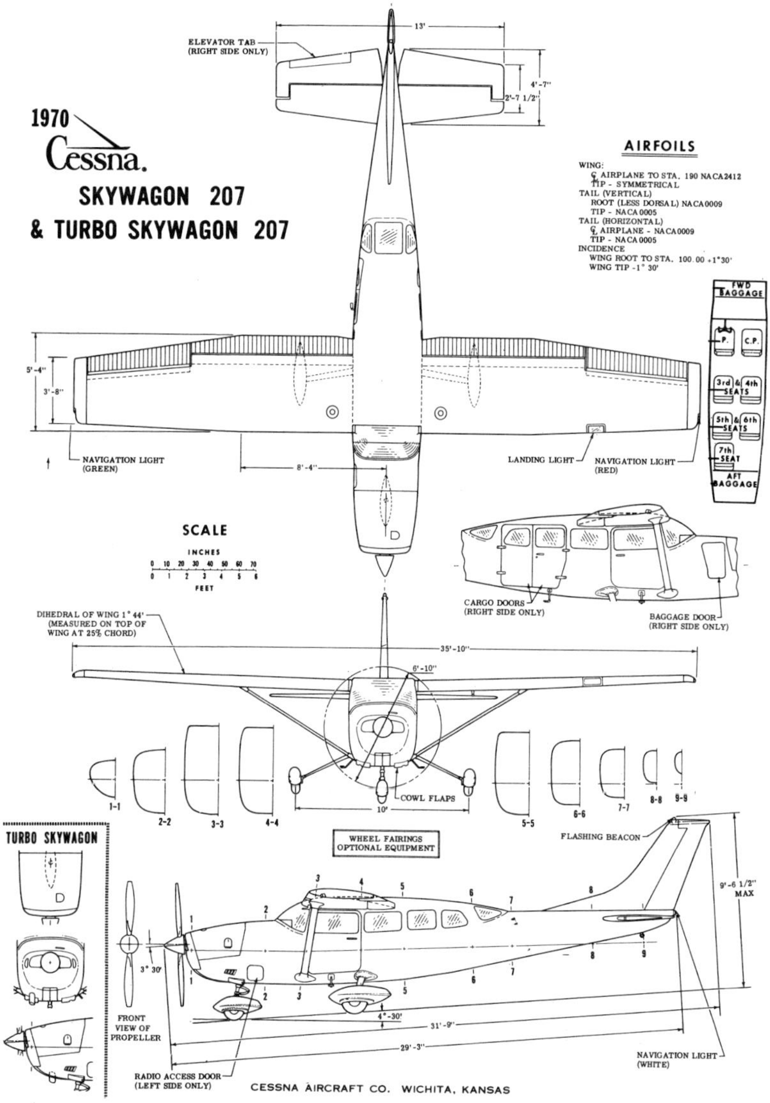
1970 Cessna.
SKYWAGON 207
& TURBO SKYWAGON 207
ELEVATOR TAB (RIGHT SIDE ONLY)
13'
4'-7"
2'-7 1/2"
AIRFOILS
WING:
℄ AIRPLANE TO STA. 190 NACA2412
TIP - SYMMETRICAL
TAIL (VERTICAL)
ROOT (LESS DORSAL) NACA0009
TIP - NACA0005
TAIL (HORIZONTAL)
℄ AIRPLANE - NACA0009
TIP - NACA0005
INCIDENCE
WING ROOT TO STA. 100.00 +1°30'
WING TIP -1° 30'
FWD BAGGAGE
P.
C.P.
3rd & 4th SEATS
5th & 6th SEATS
7th SEAT
AFT BAGGAGE
5'-4"
3'-8"
NAVIGATION LIGHT (GREEN)
8'-4"
LANDING LIGHT
NAVIGATION LIGHT (RED)
SCALE
INCHES
0 10 20 30 40 50 60 70
0 1 2 3 4 5 6
FEET
CARGO DOORS (RIGHT SIDE ONLY)
BAGGAGE DOOR (RIGHT SIDE ONLY)
DIHEDRAL OF WING 1° 44' (MEASURED ON TOP OF WING AT 25% CHORD)
35'-10"
6'-10"
COWL FLAPS
10'
1-1
2-2
3-3
4-4
5-5
6-6
7-7
8-8
9-9
WHEEL FAIRINGS OPTIONAL EQUIPMENT
TURBO SKYWAGON
FLASHING BEACON
9'-6 1/2" MAX
3° 30'
FRONT VIEW OF PROPELLER
4°-30'
31'-9"
29'-3"
NAVIGATION LIGHT (WHITE)
RADIO ACCESS DOOR (LEFT SIDE ONLY)
CESSNA AIRCRAFT CO. WICHITA, KANSAS

1970 Skywagon 206 Performance and Specifications

	LANDPLANE	FLOATPLANE	SKIPLANE
GROSS WEIGHT:	3600 lbs. 1633 kg	3500 lbs. 1588 kg	3300 lbs. 1497 kg
*SPEED: Best Power Mixture			
Top Speed at Sea Level	174 mph 280 kph	156 mph 251 kph	139 mph 224 kph
Cruise, 75% Power at 6500′	164 mph 264 kph	151 mph 243 kph	137 mph 220 kph
RANGE: Normal Lean Mixture			
Cruise, 75% Power at 6500′	650 mi. 1046 km	600 mi. 966 km	545 mi. 877 km
63 Gallons Usable Fuel	4.0 hrs.	4.0 hrs.	4.0 hrs.
No Reserve	163 mph 262 kph	150 mph 241 kph	136 mph 219 kph
Cruise, 75% Power at 6500′	830 mi. 1336 km	765 mi. 1231 km	695 mi. 1118 km
80 Gallons Usable Fuel	5.1 hrs.	5.1 hrs.	5.1 hrs.
No Reserve	163 mph 262 kph	150 mph 241 kph	136 mph 219 kph
Optimum Range at 10,000′	1020 mi. 1641 km	1005 mi. 1617 km	835 mi. 1344 km
80 Gallons Usable Fuel	7.8 hrs.	8.9 hrs.	7.9 hrs.
No Reserve	131 mph 211 kph	113 mph 182 kph	106 mph 171 kph
RATE OF CLIMB AT SEA LEVEL	920 fpm 280 mpm	855 fpm 260 mpm	800 fpm 244 mpm
SERVICE CEILING	14,800 ft. 4511 m	13,900 ft. 4237 m	11,500 ft. 3505 m
TAKEOFF:			
Ground Run (or Water Run)	900 ft. 274 m	1445 ft. 440 m	
Total Distance Over 50′ Obstacle	1780 ft. 543 m	2475 ft. 754 m	
LANDING:			
Ground Roll (or Water Run)	735 ft. 224 m	695 ft. 212 m	
Total Distance Over 50′ Obstacle	1395 ft. 425 m	1570 ft. 479 m	
STALL SPEED:			
Flaps Up, Power Off	70 mph 113 kph	67 mph 108 kph	69 mph 111 kph
Flaps Down, Power Off	61 mph 98 kph	60 mph 97 kph	60 mph 97 kph
EMPTY WEIGHT: (Approximate) (One Seat)	1710 lbs. 776 kg	2070 lbs. 939 kg	1940 lbs. 880 kg
USEFUL LOAD	1890 lbs. 857 kg	1430 lbs. 649 kg	1360 lbs. 617 kg
WING LOADING	20.7 lbs/ft.^{2} 101.0 kg/m^2	20.1 lbs/ft.^{2} 98.2 kg/m^2	19.0 lbs/ft.^{2} 92.6 kg/m^2
POWER LOADING	12.0 lbs/hp 5.4 kg/hp	11.7 lbs/hp 5.3 kg/hp	11.0 lbs/hp 5.0 kg/hp
FUEL CAPACITY: Total			
Standard	65 gal. 246 liters	65 gal. 246 liters	65 gal. 246 liters
Optional Long Range Tanks	84 gal. 318 liters	84 gal. 318 liters	84 gal. 318 liters
OIL CAPACITY: Total	12 qts. 11.4 liters	12 qts. 11.4 liters	12 qts. 11.4 liters
POWER	Six-Cylinder, Fuel-Injection Engine, 300 Rated Takeoff HP (5 min) at 2850 RPM and 285 Maximum Continuous HP at 2700 RPM.		
PROPELLER: ** Constant-Speed Two-Bladed, Diameter	82 in. 2.08 m	86 in. 2.18 m	82 in. 2.08 m
WING SPAN	35 ft. 10 in. 10.92 m	35 ft. 10 in. 10.92 m	35 ft. 10 in. 10.92 m
WING AREA	174 ft.^{2} 16.2 m^2	174 ft.^{2} 16.2 m^2	174 ft.^{2} 16.2 m^2
LENGTH	28 ft. 8.59 m	28 ft. 5 1/2 in. 8.70 m	28 ft. 8.59 m
HEIGHT: (With Depressed Nose Strut on Landplane)	9 ft. 7 1/2 in. 2.93 m	13 ft. 11 1/2 in. 4.25 m	9 ft. 7 1/2 in. 2.93 m

*Speed performance is shown for a landplane equipped with optional speed fairings which increase the speed by one MPH.

**Performance with optional three-bladed propeller is essentially the same as above.

Guide for Estimating Cost of Operation—Skywagon 206

	*PER YEAR	PER YEAR	*PER HOUR	PER HOUR	PER HOUR	PER HOUR
FLIGHT HOURS PER YEAR				300	500	700
FLIGHT MILES PER YEAR: 163 mph, 75% power at 6,500 ft., Normal Lean Mixture				48,900	81,500	114,100
GASOLINE: 15.8 gal./hr. at 46¢/gal.				$ 7.27	$ 7.27	$ 7.27
OIL: 1 pt./hr. plus oil change every 25 hrs. (12 qts. per change) at 63¢/qt.				$ .63	$ .63	$ .63
AIRPLANE & ENGINE MAINTENANCE				$ 2.15	$ 2.15	$ 2.15
TOTAL OPERATIONAL COSTS (listed above)				$10.05	$10.05	$10.05
RESERVE FOR ENGINE OVERHAUL				$ 2.13	$ 2.13	$ 2.13
STORAGE: at $50.00 per month		$ 600		$ 2.00	$ 1.20	$ .86
INSURANCE (a guide for the following coverage): All risks hull coverage (Deductibles: In motion $250.00) (Not in motion $50.00)		$1,544				
Liability Coverage (Single limit $250,000) (Including passengers)		$ 140				
TOTAL INSURANCE (Industrial private business & pleasure use by a qualified pilot)		$1,684		$ 5.61	$ 3.37	$ 2.41
COST PER HOUR				$19.79	$16.75	$15.45
COST PER MILE: 75% power, 163 mph				$ .121	$ .103	$ .095
COST PER SEAT MILE: 6 People				$ .020	$ .017	$ .016

FOR DEPRECIATION: Consult your tax authority. To estimate the per hour or per mile cost, divide annual depreciation amount by annual hours or miles.

*YOUR COSTS: No doubt vary from these national averages. This column is included for you to compute your estimate of your hours, miles and costs.

1970 Turbo Skywagon 206 Performance and Specifications

	LANDPLANE	FLOATPLANE	SKIPLANE
GROSS WEIGHT	3600 lbs. 1633 kg	3600 lbs. 1633 kg	3300 lbs. 1497 kg
*SPEED: Best Power Mixture			
Top Speed at 19,000 ft.	200 mph 322 kph	181 mph 291 kph	167 mph 269 kph
**Cruise, 75% Power at 24,000′	184 mph 296 kph	164 mph 270 kph	152 mph 245 kph
Cruise, 75% Power at 10,000′	170 mph 272 kph	153 mph 246 kph	141 mph 227 kph
RANGE: Normal Lean Mixture			
**Cruise, 75% Power at 24,000′	700 mi. 1127 km	620 mi. 1080 km	570 mi. 917 km
63 Gallons Usable Fuel	3.8 hrs.	3.8 hrs.	3.8 hrs.
No Reserve	182 mph 293 kph	162 mph 269 kph	151 mph 243 kph
Cruise, 75% Power at 10,000′	645 mi. 1038 km	585 mi. 941 km	530 mi. 853 km
63 Gallons Usable Fuel	3.8 hrs.	3.8 hrs.	3.8 hrs.
No Reserve	168 mph 270 kph	152 mph 245 kph	140 mph 225 kph
**Cruise, 75% Power at 24,000′	890 mi. 1432 km	790 mi. 1311 km	740 mi. 1191 km
80 Gallons Usable Fuel	4.9 hrs.	4.9 hrs.	4.9 hrs.
No Reserve	182 mph 293 kph	162 mph 269 kph	151 mph 243 kph
Cruise, 75% Power at 10,000′	820 mi. 1320 km	745 mi. 1199 km	680 mi. 1094 km
80 Gallons Usable Fuel	4.9 hrs.	4.9 hrs.	4.9 hrs.
No Reserve	168 mph 270 kph	152 mph 245 kph	140 mph 225 kph
Optimum Range at 15,000′	1050 mi. 1690 km	885 mi. 1424 km	810 mi. 1304 km
80 Gallons Usable Fuel	7.6 hrs.	6.9 hrs.	6.9 hrs.
No Reserve	139 mph 224 kph	128 mph 206 kph	118 mph 190 kph
RATE OF CLIMB AT SEA LEVEL	1030 fpm 314 mpm	950 fpm 290 mpm	920 fpm 280 mpm
SERVICE CEILING	26,300 ft. 8020 m	24,200 ft. 7375 m	23,500 ft. 7163 m
TAKEOFF:			
Ground Run (or Water Run)	910 ft. 277 m	1400 ft. 427 m	
Total Distance Over 50′ Obstacle	1810 ft. 552 m	2400 ft. 731 m	
LANDING:			
Ground Roll (or Water Run)	735 ft. 224 m	710 ft. 216 m	
Total Distance Over 50′ Obstacle	1395 ft. 425 m	1610 ft. 491 m	
STALL SPEED:			
Flaps Up, Power Off	70 mph 113 kph	68 mph 109 kph	69 mph 111 kph
Flaps Down, Power Off	61 mph 98 kph	61 mph 98 kph	60 mph 98 kph
EMPTY WEIGHT: (Approximate) (One-Seat)	1810 lbs. 821 kg	2170 lbs. 984 kg	2040 lbs. 925 kg
USEFUL LOAD:	1790 lbs. 812 kg	1430 lbs. 649 kg	1260 lbs. 572 kg
WING LOADING	20.7 lbs/ft.^{2} 101.0 kg/m^2	20.7 lbs/ft.^{2} 101.0 kg/m^2	19.0 lbs/ft.^{2} 92.6 kg/m^2
POWER LOADING	12.6 lbs/hp 5.7 kg/hp	12.6 lbs/hp 5.7 kg/hp	11.6 lbs/hp 5.3 kg/hp
FUEL CAPACITY: Total			
Standard	65 gal. 246 liters	65 gal. 246 liters	65 gal. 246 liters
Optional Long Range Tanks	84 gal. 318 liters	84 gal. 318 liters	84 gal. 318 liters
OIL CAPACITY: Total	13 qts. 12.3 liters	13 qts. 12.3 liters	13 qts. 12.3 liters
POWER	Six-Cylinder, Turbo-Super charged, Fuel-Injection Engine, 285 Rated HP at 2700 RPM and 32.5 inches M.P.		
PROPELLER: *** Constant-Speed, Two-Bladed, Diameter	82″ 2.08 m	82″ 2.08 m	86″ 2.18 m
WING SPAN	35′ 10 in. 10.92 m	35′ 10 in. 10.92 m	35′ 10 in. 10.92 m
WING AREA:	174′^{2} 16.2 m^2	174′^{2} 16.2 m^2	174′^{2} 16.2 m^2
LENGTH	28′ 8.59 m	28′ 8.59 m	28′ 5 1/2 in. 8.70 m
HEIGHT: (With Depressed Nose Strut on Landplane)	9′ 7 1/2 in. 2.93 m	9′ 7 1/2 in. 2.93 m	13′ 11 1/2 in. 4.25 m

*Speed performance is shown for a landplane equipped with optional speed fairings which increase the speed by one MPH.

**Replace 24,000 feet with 20,000 feet for floatplane and skiplane.

***Performance with optional three-bladed propeller is essentially the same as above.

Guide for Estimating Cost of Operation—Turbo Skywagon 206

	*PER YEAR	PER YEAR	*PER HOUR	PER HOUR	PER HOUR	PER HOUR
FLIGHT HOURS PER YEAR				300	500	700
FLIGHT MILES PER YEAR: 168 mph, 75% power at 10,000 ft., Normal Lean Mixture.				50,400	84,000	117,600
GASOLINE: 16.6 gal./hr. at 46¢/gal.				$ 7.64	$ 7.64	$ 7.64
OIL: 1 pt./hr. plus oil change every 50 hrs. (13 qts. per change) at 63/qt.				$ .49	$ .49	$ .49
AIRPLANE & ENGINE MAINTENANCE				$ 2.40	$ 2.40	$ 2.40
TOTAL OPERATIONAL COSTS (listed above)				$10.53	$10.53	$10.53
RESERVE FOR ENGINE OVERHAUL				$ 2.57	$ 2.57	$ 2.57
STORAGE: at $50.00 per month		$ 600		$ 2.00	$ 1.20	$ .86
INSURANCE (a guide for the following coverage) All risks hull coverage (Deductibles: In motion $250.00) (Not in motion $50.00)		$1,739				
Liability Coverage (Single limit $250,000) (Including passengers)		$ 140				
TOTAL INSURANCE (Industrial private business & pleasure use by a qualified pilot)		$1,879		$ 6.26	$ 3.76	$ 2.68
COST PER HOUR				$21.36	$18.06	$16.64
COST PER MILE: 75% power, 168 mph				$ .127	$ .108	$ .099
COST PER SEAT MILE: 6 People				$ .021	$ .018	$ .017

FOR DEPRECIATION: Consult your tax authority. To estimate the per hour or per mile cost, divide annual depreciation amount by annual hours or miles.

*YOUR COSTS: No doubt vary from these national averages. This column is included for you to compute your estimate of your hours, miles and costs.

1970 Skywagon 207 Specifications and Performance

GROSS WEIGHT:	3800 lbs, 1724 kg
*SPEED: Best Power Mixture	
Top Speed at Sea Level	168 mph, 270 kph
Cruise, 75% Power at 6500′	159 mph, 256 kph
RANGE: Normal Lean Mixture	
Cruise, 75% Power at 6500′	585 mi, 941 km
58 Gallons usable fuel	3.7 hrs
No Reserve	158 mph, 254 kph
Cruise, 75% Power at 6500′	775 mi, 1247 km
77 Gallons usable fuel	4.9 hrs
No Reserve	158 mph, 254 kph
Optimum Range at 10,000′	925 mi, 1489 km
77 Gallons usable fuel	7.0 hrs
No Reserve	131 mph, 211 kph
RATE OF CLIMB AT SEA LEVEL:	810 fpm, 247 mpm
SERVICE CEILING:	13,300 ft, 4054 m
TAKEOFF:	
Ground Run	1100 ft, 335 m
Total Distance Over 50' Obstacle	1970 ft, 600 m
LANDING:	
Ground Roll	765 ft, 233 m
Total Distance Over 50' Obstacle	1500 ft, 457 m
STALL SPEED:	
Flaps Up, Power Off	75 mph, 121 kph
Flaps Down, Power Off	67 mph, 108 kph
EMPTY WEIGHT: (Approximate)	1880 lbs, 853 kg
USEFUL LOAD:	1920 lbs, 871 kg
WING LOADING:	21.8 lbs/ft^2, 106.7 kg/m^2
POWER LOADING:	12.7 lbs/hp, 5.7 kg/hp
FUEL CAPACITY: Total	
Standard	65 gal, 246 liters
Optional Long Range Tanks	84 gal, 318 liters
OIL CAPACITY: Total	12 qts, 11.4 liters
POWER:	Six-Cylinder, Fuel Injection Engine 300 Rated Takeoff HP (5 min) at 2850 RPM and 285 Maximum continuous HP at 2700 RPM
**PROPELLER: Two-Blade, Constant-Speed; Diameter	82 in, 2.08 m
WING SPAN:	35 ft 10 in, 10.92 m
WING AREA:	174 ft^2, 16.2 m^2
LENGTH:	31 ft 9 in, 9.68 m
HEIGHT: (With Depressed Nose Strut)	9 ft 6-1/2 in, 2.91 m

* Speed performance is shown for an airplane equipped with optional speed fairings, which increase the speed by one MPH

** Optional 3-blade propeller available. Performance with 3-blade propeller is essentially the same as above.

1970 Turbo Skywagon 207 Specifications and Performance

GROSS WEIGHT:	3800 lbs,1724 kg
*SPEED: Best Power Mixture	
Top Speed at 17,000′Ft.	189 mph, 304 kph
Cruise, 75% Power at 20,000′	176 mph, 283 kph
Cruise, 75% Power at 10,000′	163 mph, 262 kph
RANGE: Normal Lean Mixture	
Cruise, 75% Power at 20,000′	610 mi, 982 km
58 Gallons usable fuel	3.5 hrs
No Reserve	174 mph, 280 kph
Cruise, 75% Power at 10,000′	565 mi, 909 km
58 Gallons usable fuel	3.5 hrs
No Reserve	161 mph, 259 kph
Cruise, 75% Power at 20,000′	820 mi, 1320 km
77 Gallons usable fuel	4.7 hrs
No Reserve	174 mph, 280 kph
Cruise, 75% Power at 10,000′	755 mi, 1215 km
77 Gallons usable fuel	4.7 hrs
No Reserve	161 mph, 259 kph
Optimum Range at 20,000′	910 mi, 1445 km
77 Gallons usable fuel	5.8 hrs
No Reserve	156 mph, 251 kph
RATE OF CLIMB AT SEA LEVEL:	885 fpm, 270 mpm
SERVICE CEILING:	24,200 ft, 7376 m
TAKEOFF:	
Ground Run	1100 ft, 335 m
Total Distance Over 50' Obstacle	1970 ft, 600 m
LANDING:	
Ground Roll	765 ft, 233 m
Total Distance Over 50' Obstacle	1500 ft, 457 m
STALL SPEED:	
Flaps Up, Power Off	75 mph, 121 kph
Flaps Down, Power Off	67 mph, 108 kph
EMPTY WEIGHT: (Approximate)	1980 lbs, 898 kg
USEFUL LOAD:	1820 lbs, 826 kg
WING LOADING:	21.8 lbs/ft^2, 106.7 kg/m^2
POWER LOADING:	12.7 lbs/hp, 5.7 kg/hp
FUEL CAPACITY: Total	
Standard	65 gal, 246 liters
Optional Long Range Tanks	84 gal, 318 liters
OIL CAPACITY: Total	13 qts, 12.3 liters
POWER:	Six-Cylinder, Turbo-Charged Engine 300 Rated Takeoff HP (5 min) at 2700 RPM and 285 Maximum Continuous HP at 2600 RPM
**PROPELLER: Two-Blade, Constant-Speed; Diameter	81.5 in, 2.07 m
WING SPAN:	35 ft 10 in, 10.92 m
WING AREA:	174 ft^2, 16.2 m^2
LENGTH:	31 ft 9 in, 9.68 m
HEIGHT: (With Depressed Nose Strut)	9 ft 6-1/2 in, 2.91 m

* Speed performance is shown for an airplane equipped with optional speed fairings, which increase the speed by one MPH.

** Optional 3-blade propeller available. Performance with 3-blade propeller is essentially the same as above.

Guide for Estimating Cost of Operation—Skywagon 207

	*PER YEAR	PER YEAR	*PER HOUR	PER HOUR	PER HOUR	PER HOUR
FLIGHT HOURS PER YEAR				300	500	700
FLIGHT MILES PER YEAR: 158 mph, 75% power at 6,500 ft., Normal Lean Mixture				47,400	79,000	110,600
GASOLINE: 15.7 gal./hr. at 46¢/gal				$ 7.22	$ 7.22	$ 7.22
OIL: 1 pt./hr. plus oil change every 25 hrs. (12 qts. per change) at 63¢/qt				$ .63	$ .63	$ .63
AIRPLANE & ENGINE MAINTENANCE				$ 2.15	$ 2.15	$ 2.15
TOTAL OPERATIONAL COSTS (listed above)				$10.00	$10.00	$10.00
RESERVE FOR ENGINE OVERHAUL				$ 2.13	$ 2.13	$ 2.13
STORAGE: at $50.00 per month		$ 600		$ 2.00	$ 1.20	$ .86
INSURANCE (a guide for the following coverage) All risks hull coverage (Deductibles: In motion $250.00) (Not in motion $50.00)		$1,600				
Liability Coverage (Single limit $250,000) (Including passengers)		$ 140				
TOTAL INSURANCE (Industrial private business & pleasure use by a qualified pilot)		$1,740		$ 5.80	$ 3.48	$ 2.49
COST PER HOUR				$19.93	$16.81	$15.48
COST PER MILE: 75% power, 158 mph				$ .126	$ .106	$ .098
COST PER SEAT MILE: 7 People				$ .018	$ .015	$ .014

FOR DEPRECIATION: Consult your tax authority. To estimate the per hour or per mile cost, divide annual depreciation amount by annual hours or miles.

*YOUR COSTS: No doubt vary from these national averages. This column is included for you to compute your estimate of your hours, miles and costs.

Guide for Estimating Cost of Operation—Turbo Skywagon 207

	*PER YEAR	PER YEAR	*PER HOUR	PER HOUR	PER HOUR	PER HOUR
FLIGHT HOURS PER YEAR..................				300	500	700
FLIGHT MILES PER YEAR: 161 mph, 75% power at 10,000 ft., Normal Lean Mixture.				48,300	80,500	112,700
GASOLINE: 16.6 gal./hr. at 46¢/gal...........				$ 7.64	$ 7.64	$ 7.64
OIL: 1 pt./hr. plus oil change every 50 hrs. (13 qts. per change) at 63¢/qt............				$ .49	$ .49	$ .49
AIRPLANE & ENGINE MAINTENANCE.........				$ 2.40	$ 2.40	$ 2.40
TOTAL OPERATIONAL COSTS (listed above)....				$10.53	$10.53	$10.53
RESERVE FOR ENGINE OVERHAUL...........				$ 2.84	$ 2.84	$ 2.84
STORAGE: at $50.00 per month.............		$ 600		$ 2.00	$ 1.20	$ .86
INSURANCE (a guide for the following coverage) All risks hull coverage (Deductibles: In motion $250.00) (Not in motion $50.00)....................		$1,790				
Liability Coverage (Single limit $250,000) (Including passengers)		$ 140				
TOTAL INSURANCE (Industrial private business & pleasure use by a qualified pilot)....................		$1,930		$ 6.43	$ 3.86	$ 2.76
COST PER HOUR.........................				$21.80	$18.43	$16.99
COST PER MILE: 75% power, 161 mph.......				$ .135	$ .114	$ .106
COST PER SEAT MILE: 7 People.............				$ .019	$ .016	$ .015

FOR DEPRECIATION: Consult your tax authority. To estimate the per hour or per mile cost, divide annual depreciation amount by annual hours or miles.

*YOUR COSTS: No doubt vary from these national averages. This column is included for you to compute your estimate of your hours, miles and costs.

CHAPTER 10: CESSNA 210/CENTURION

The first Cessna 210 was the 1960 Model, announced late in 1959. At that time, many of us tended to think of it as a slicked-up 182 with a crazy retract gear and bigger engine (IO-470E fuel-injected Continental of 260-hp), but it wasn't. It was an all-new design aimed at the Bonanza/Comanche market. It had a top speed of 199 mph, a cruise (75% @ 7,000) of 190 mph and should have been right on target; but customer acceptance was slow. Ten years later, sales of this craft still lagged behind Cessna's expectations and the explanation remained obscure, probably clouded by an un-voiced (and un-reasoned) customer prejudice – perhaps something some people didn't like about the plane and were ashamed or reluctant to admit.

We're guessing about this, because we don't know why Cessna's top performer in the single-engine class has not taken a bigger share of that market. From the beginning, it has performed just as Cessna claimed, and has never been accused of possessing any vicious traits.

Anyway, beginning with the 1969 Model, the question – if it remains – will have to come from the other side of our mouth, because the new Centurions (nee 210) are different airplanes and demand fresh evaluations.

Known as the "210" during its first four years of production, this design became the "Centurion" in 1964 when it received a re-designed wing with more flap area, four additional inches on the tail surfaces and the IO-520A 285-hp engine (turtle deck with back windows was added in 1962, along with four extra inches of cabin width). In 1967, the Centurion appeared with a full-cantilever wing; and in 1968 de-icing and anti-icing systems for prop, wing and tail were offered.

But the strutless Centurions of 1967-68, though structurally sound, apparently didn't look that way to many because of their pronounced (three degrees) dihedral. At least, that was Cessna's conclusion. Therefore, the cantilever wing was flattened-out on the 1969 Model, with only one degree of dihedral remaining, and the

1970 Cessna Turbo Centurion

1970 Cessna Centurion

ailerons inter-connected with the rudder via a spring load to keep good stability.

Then, in 1970, the Centurion received a 25% bigger cabin to make it a full six-place airplane, with new side windows extending almost half the length of the cabin sans posts. Also in 1970, the Centurion went to the tubular-steel main-gear legs for softer landings and a wider stance (the nose-wheel strut was shortened on the 1969 Model, and the "chin" housing eliminated).

The new gear retracts completely into the fuselage within a few seconds at the touch of a control on the pedestal. The system has a fluid reservoir and is under pressure only during the time the gear is being activated. This improves reliability and should reduce service costs.

The cabin is accessible through a yard-wide door on each side, and offers a full assortment of convenience items such as individual air vents, ash trays, cigarette lighter and magazine pockets – all close at hand. Other standard cabin accessories include sun visors, map pockets, arm rests, an integrally-mounted overhead light console for the pilot's area and individual lights above the passenger seats.

Some of the purists may complain about the inter-connected roll and yaw controls, but this objection won't be as "pure" as the gung-ho crowd will have you believe, because this principle was employed by the Wright brothers on the world's first successful airplane. It seems to us that the only pilots who have a valid objection to inter-connected controls are the aerobatic types; and Cessna clearly didn't intend that anyone should slow-roll the Centurion across country at 200 miles per. However, you can over-ride the spring-load inter-connect anytime you want or need to.

If you've been flying a Super Skylane or anything of comparable muscle, transition to the Centurion will be easy. But if you're accustomed to smaller and slower craft, you really should have at least three hours of dual (and a minimum of 150 hours in your logbook) before sashaying off to far away places in this – or any other – high performance retractable. It is a big airplane, and a sudden one. Cruising at 185-190 mph requires that you think more than one check-point ahead in matters of fuel management, efficient let-downs and weather evaluation. In short, it's a lot more airplane than a Skyhawk, and it demands a higher level of competence to safely operate it.

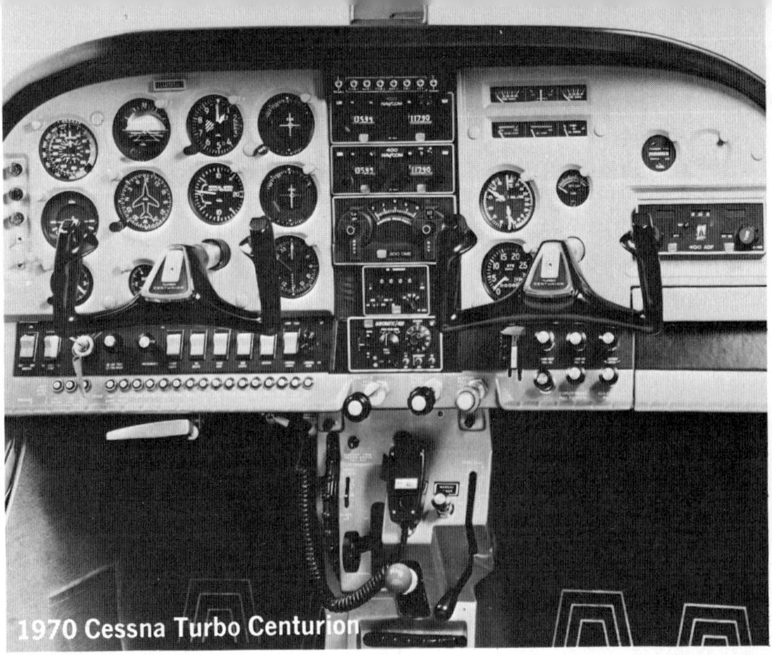

We don't mean by this that the Centurion is physically hard to fly. It is not. It is very stable, with a solid, big-airplane feel, as a long-legged personal transport should be. What we mean by "higher level of competence" is that, as aircraft weight and speed goes up, so does the penalty for error.

The Turbo Centurion is of course the high-altitude performer, and the one that makes a "standard day" out of a hot and humid one. The turbo system assures maximum horsepower up to 19,000 feet, and beyond that keeps working to compress the thin air into a richer source of oxygen for the engine. This system adds no extra controls to adjust or gauges to monitor, and no constant adjusting of the throttle is required during the climb to 19,000. The turbine is automatically regulated by an absolute pressure controller and an over-boost control valve which automatically bleeds off excessive compressor discharge pressure. And all this is housed in a (wouldn't you just know it?) "Jet-Flow" cowling.

The Turbo Centurion comes with a 74-cubic-foot capacity oxygen system as standard equipment. This system includes an overhead oxygen console with a capacity gauge system control, control wheel microphone switch and automatic regulator.

Suggested list price of the 1970 Model Centurion was $34,725 while the Turbo Centurion was priced at $39,250.

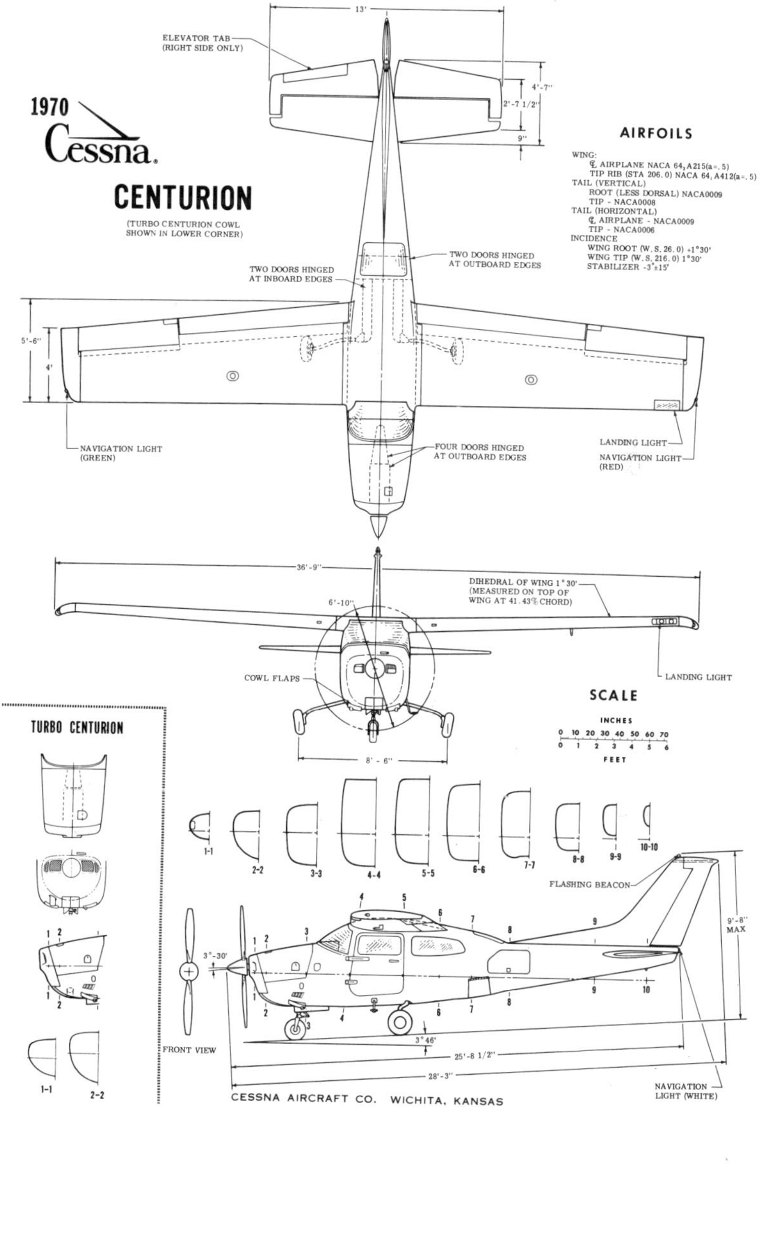
1970 Cessna
CENTURION
(TURBO CENTURION COWL SHOWN IN LOWER CORNER)
13'
ELEVATOR TAB (RIGHT SIDE ONLY)
4'-7"
2'-7 1/2"
9"
AIRFOILS
WING:
℄ AIRPLANE NACA 64,A215(a=.5)
TIP RIB (STA 206.0) NACA 64,A412(a=.5)
TAIL (VERTICAL)
ROOT (LESS DORSAL) NACA0009
TIP - NACA0008
TAIL (HORIZONTAL)
℄ AIRPLANE - NACA0009
TIP - NACA0006
INCIDENCE
WING ROOT (W.S. 26.0) +1°30'
WING TIP (W.S. 216.0) 1°30'
STABILIZER -3°±15'
TWO DOORS HINGED AT OUTBOARD EDGES
TWO DOORS HINGED AT INBOARD EDGES
5'-6"
4'
NAVIGATION LIGHT (GREEN)
FOUR DOORS HINGED AT OUTBOARD EDGES
LANDING LIGHT
NAVIGATION LIGHT (RED)
36'-9"
DIHEDRAL OF WING 1°30' (MEASURED ON TOP OF WING AT 41.43% CHORD)
6'-10"
COWL FLAPS
LANDING LIGHT
SCALE
INCHES
0 10 20 30 40 50 60 70
0 1 2 3 4 5 6
FEET
8' - 6"
TURBO CENTURION
1-1
2-2
3-3
4-4
5-5
6-6
7-7
8-8
9-9
10-10
FLASHING BEACON
9'-8" MAX
3°-30'
3° 46'
25'-8 1/2"
28'-3"
FRONT VIEW
NAVIGATION LIGHT (WHITE)
CESSNA AIRCRAFT CO. WICHITA, KANSAS

1970 Cessna Centurion Performance and Specifications

GROSS WEIGHT:	3800 lbs.	1724 kg
SPEED: Best Power Mixture		
Top Speed at Sea Level	200 mph	322 kph
Cruise, 75% Power at 7500 ft.	188 mph	303 kph
RANGE: Normal Lean Mixture		
Cruise, 75% Power at 7500 ft. 64 Usable Gallons, No Reserve	765 mi 4.1 hrs 187 mph	1231 km 4.1 hrs 301 kph
Cruise, 75% Power at 7500 ft. 89 Usable Gallons, No Reserve	1065 mi 5.7 hrs 187 mph	1714 km 5.7 hrs 301 kph
Optimum Range at 10,000 ft. 89 Usable Gallons, No Reserve	1250 mi 8.1 hrs 154 mph	2012 km 8.1 hrs 248 kph
RATE OF CLIMB AT SEA LEVEL:	860 fpm	262 mpm
SERVICE CEILING:	15,500 ft.	4755 m
TAKEOFF:		
Ground Run	1100 ft.	335 m
Total Distance Over 50 ft. Obstacle	1900 ft.	579 m
LANDING:		
Ground Roll	765 ft.	233 m
Total Distance Over 50 ft. Obstacle	1500 ft.	457 m
STALL SPEED:		
Flaps Up, Power Off	75 mph	121 kph
Flaps Down, Power Off	65 mph	105 kph
EMPTY WEIGHT: (Approximate)	2080 lbs.	944 kg
USEFUL LOAD:	1720 lbs.	780 kg
WING LOADING:	21.6 lbs/ft.^{2}	105.4 kg/m^2
POWER LOADING:	12.7 lbs/hp	5.7 kg/hp
FUEL CAPACITY: Total	90 gal.	341 liters
OIL CAPACITY: Total	10 qts.	9.5 liters
POWER:	Six-Cylinder, Fuel Injection Engine, 300 Rated Takeoff HP (5 min) at 2850 RPM and 285 Maximum Continuous HP at 2700 RPM	
*PROPELLER: Two-Bladed, Constant Speed; Diameter	82 in.	2.08 m
WING SPAN:	36 ft. 9 in.	11.2 m
WING AREA:	176 ft.^{2}	16.4 m^2
LENGTH:	28 ft. 3 in.	8.61 m
HEIGHT: (With Depressed Nose Strut)	9 ft. 8 in.	2.95 m

* Performance for optional 3-bladed propeller is essentially the same as above.

Guide for Estimating Cost of Operation—Centurion

	*PER YEAR	PER YEAR	*PER HOUR	PER HOUR	PER HOUR	PER HOUR
FLIGHT HOURS PER YEAR				300	500	700
FLIGHT MILES PER YEAR: 187 mph, 75% power at 6,500 ft., Normal Lean Mixture				56,100	93,500	130,900
GASOLINE: 15.6 gal./hr. at 46¢/gal.				$ 7.18	$ 7.18	$ 7.18
OIL: 1 pt./hr. plus oil change every 25 hrs. (12 qts. per change) at 63¢/qt.				$ .63	$ .63	$ .63
AIRPLANE & ENGINE MAINTENANCE				$ 3.20	$ 3.20	$ 3.20
TOTAL OPERATIONAL COSTS (listed above)				$11.01	$11.01	$11.01
RESERVE FOR ENGINE OVERHAUL				$ 2.18	$ 2.18	$ 2.18
STORAGE: at $50.00 per month		$ 600		$ 2.00	$ 1.20	$ .86
INSURANCE (a guide for the following coverage): All risks hull coverage (Deductibles: In motion $250.00) (Not in motion $50.00)		$1,175				
Liability Coverage (Single limit $250,000) (Including passengers)		$ 140				
TOTAL INSURANCE (Industrial private business & pleasure use by a qualified pilot)		$1,315		$ 4.38	$ 2.63	$ 1.88
COST PER HOUR				$19.57	$17.02	$15.93
COST PER MILE: 75% power, 187 mph				$ .105	$ .091	$ .085
COST PER SEAT MILE: 6 People				$ .018	$ .016	$ .014

FOR DEPRECIATION: Consult your tax authority. To estimate the per hour or per mile cost, divide annual depreciation amount by annual hours or miles.

*YOUR COSTS: No doubt vary from these national averages. This column is included for you to compute your estimate of your hours, miles and costs.

1970 Turbo Centurion Performance and Specifications

GROSS WEIGHT:	3800 lbs.	1724 kg
SPEED: Best Power Mixture		
Top Speed at 19,000 ft.	230 mph	370 kph
Cruise, 75% Power at 24,000 ft.	219 mph	352 kph
Cruise, 75% Power at 10,000 ft.	195 mph	314 kph
RANGE: Normal Lean Mixture		
Cruise, 75% Power at 24,000 ft. 64 Usable Gallons, No Reserve	845 mi 3.9 hrs 217 mph	1360 km 3.9 hrs 349 kph
Cruise, 75% Power at 10,000 ft. 64 Usable Gallons, No Reserve	755 mi 3.9 hrs 193 mph	1215 km 3.9 hrs 311 kph
Cruise, 75% Power at 24,000 ft. 89 Usable Gallons, No Reserve	1170 mi 5.4 hrs 217 mph	1883 km 5.4 hrs 349 kph
Cruise, 75% Power at 10,000 ft. 89 Usable Gallons, No Reserve	1040 mi 5.4 hrs 193 mph	1673 km 5.4 hrs 311 kph
Optimum Range at 24,000 ft. 89 Gallons, No Reserve	1330 mi 7.6 hrs 175 mph	2140 km 7.6 hrs 282 kph
RATE OF CLIMB AT SEA LEVEL:	930 fpm	283 mpm
SERVICE CEILING:	28,500 ft.	8687 m
TAKEOFF:		
Ground Run	1170 ft.	357 m
Total Distance Over 50 ft. Obstacle	2030 ft.	619 m
LANDING:		
Ground Roll	765 ft.	233 m
Total Distance Over 50 ft. Obstacle	1500 ft.	457 m
STALL SPEED:		
Flaps Up, Power Off	75 mph	121 kph
Flaps Down, Power Off	65 mph	105 kph
EMPTY WEIGHT: (Approximate)	2180 lbs	989 kg
USEFUL LOAD:	1620 lbs	735 kg
WING LOADING:	21.6 lbs/ft.^{2}	105.4 kg/m^2
POWER LOADING:	13.3 lbs/hp	6.0 kg/hp
FUEL CAPACITY: Total	90 gal.	341 liters
OIL CAPACITY: Total	11 qts.	10.4 liters
POWER: Six-Cylinder, Turbocharged Fuel Injection Engine, 285 Rated HP at 2700 RPM		
*PROPELLER: Two-Bladed, Constant Speed, Diameter	82 in.	2.08 m
WING SPAN:	36 ft. 9 in.	11.2 m
WING AREA:	176 ft.^{2}	16.4 m^2
LENGTH:	28 ft. 3 in.	8.61 m
HEIGHT: (With Depressed Nose Strut)	9 ft. 8 in.	2.95 m

* Performance for optional 3-bladed propeller is essentially the same as above.

Guide for Estimating Cost of Operation—Turbo Centurion

	•PER YEAR	PER YEAR	•PER HOUR	PER HOUR	PER HOUR	PER HOUR
FLIGHT HOURS PER YEAR.				300	500	700
FLIGHT MILES PER YEAR: 193 mph, 75% power at 10,000 ft., Normal Lean Mixture. .				57,900	96,500	135,100
GASOLINE: 16.4 gal./hr. at 46¢/gal.				$ 7.54	$ 7.54	$ 7.54
OIL: 1 pt./hr. plus oil change every 50 hrs. (13 qts. per change) at 63¢/qt.				$ .49	$ .49	$ .49
AIRPLANE & ENGINE MAINTENANCE.				$ 3.45	$ 3.45	$ 3.45
TOTAL OPERATIONAL COSTS (listed above).				$11.48	$11.48	$11.48
RESERVE FOR ENGINE OVERHAUL.				$ 2.73	$ 2.73	$ 2.73
STORAGE: at $50.00 per month.		$ 600		$ 2.00	$ 1.20	$ .86
INSURANCE (a guide for the following coverage): All risks hull coverage (Deductibles: In motion $250.00) (Not in motion: $50.00).		$1,327				
Liability Coverage (Single limit $250,000) (Including passengers).		$ 140				
TOTAL INSURANCE (Industrial private business & pleasure use by a qualified pilot)		$1,467		$ 4.89	$ 2.93	$ 2.10
COST PER HOUR.				$21.10	$18.34	$17.17
COST PER MILE: 75% power, 193 mph.				$ .109	$ .095	$ .089
COST PER SEAT MILE: 6 People.				$ .018	$ .016	$ .015

FOR DEPRECIATION: Consult your tax authority. To estimate the per hour or per mile cost, divide annual depreciation amount by annual hours or miles.

•YOUR COSTS: No doubt vary from these national averages. This column is included for you to compute your estimate of your hours, miles and costs.

Cessna Commercial Airplane Sales

SINGLE-ENGINE MODEL	1946-47	1948	1949	1950	1951	1952	1953	1954	1955	1956	1957	1958	1959	1960	1961	1962	1963	1964	1965	1966	1967	1968	1969	Total
120	2039	116	9																					2,164
140	4376	603	205	318	38		2																	5,560
190	8	45	32	58	28	26	7																	204
195	61	153	142	121	69	161	116	44	16	6	1													890
170		714	469	637	416	1186	663	495	448	72	36													5,136
Skywagon 180							644	646	880	530	438	262	258	283	130	122	128	146	157	167	90	97	110	5,088
172/Skyhawk									173	1419	939	790	874	1015	903	889	1146	1401	1436	1597	839	1206	1170	15,797
182/Skylane										983	835	874	774	667	575	824	642	778	865	993	836	778	673	11,097
175/Skylark												702	727	501	126	50	13							2,119
150												122	648	354	344	331	472	804	1637	3087	2114	2007	1714	13,634
F150 (French-built)																				100	150	152	136	538
F172 (French-built)																			114	108	142	103	84	551
210													43	610	171	281	156	283	224	257	226	190	181	2,622
Skywagon 185															293	229	157	116	181	193	151	125	167	1,612
205																165	353	55	1					574
Super Skylane																		41	126	161	106	88	60	582
Skywagon 206																	61	240	181	252	243	229	207	1,413
Agwagon																				193	95	143	141	572
177/Cardinal																					557	601	255	1,413
Reims Rocket (French-built)																					1	65	81	147
Skywagon 207																							123	123
MULTI-ENGINE MODEL																								
310								14	228	223	150	176	240	290	136	169	128	197	217	281	174	208	201	3,032
Skynight															68	64	60	72	91	127	68	24	1	575
Skymaster																	140	55						195
411																			122	115	41	22	1	301
Super Skymaster																			227	238	226	207	199	1,097
401																				16	68	102	122	308
402																				3	66	68	114	251
414																							19	19
421							*	* *	**												40	163	128	331
TOTALS	6,484	1631	857	1134	551	1373	1434	1200	1746	3235	2399	2926	3564	3720	2746	3124	3456	4188	5629	7888	6233	6578	5887	77,945

Commercial production prior to 1941	377
Military production 1940 - 1969	11,480
Miscellaneous production	1,095
GRAND TOTAL	90,897

*Totals include following: 1953 - 2 Model 305A; 1954 - 1 Model 305A; 1955 - 1 Model 305A; 1956 - 2 Model 325

Cessna Commercial Production Prior to 1947

Year of Manufacture	Model	Total Built
1911	First Cessna	1
1912	Silver Wings	1
1913-1917	1-place monoplanes	6
1926	4-Olace monoplane	1
1928	AA	13
1928	AF	2
1928	AS	4
1928-1930	AW	48
1928	BW	13
1929	AAX	1
1929	AC	1
1929	CW-6	1
1929	Goebel special	1
1929	CM-1 racer	1
1929	DC-6	1
1929-30 & 34	DC-6A and B	85
1930	Baby Cessna	1
1930	EC-1	1
1930	FC-1	1
1930	GC-1 racer	1
1930	GC-2 racer	1
1932	EC-2	1
1932	CR-1 racer	1
1933	CR-2 racer	1
1933	CR-3 racer	1
1933	C-3	1
1935-1936	C-34	43
1937	C-37	46
1938	C-38	15
1938-1941	C-145	41
1938-1941	C-165	42
	Total	377

Cessna Military Production 1940—1969

Year of Manufacture	Model	Total Built
1940-1944	T-50*	5,402
1950-1952	LC-126	83
1950-1959 & 1962-63	O-1 (L-19)	3,431
1957-1958	U-3A	160
1961	U-3B	36
1955	OE-2	25
1954-1969	T-37	1,176
1968-1969	A-37B	180
1966-1967	T-41A	204
1967	T-41B	255
1967-1969	T-41C	45
1968	T-41D	86
1967-1969	O-2A, O-2B	397
		11,480

*Includes AT-8; AT-17; UC-78 and Crane series

Cessna Miscellaneous Production

Year of Manufacture	Model	Total Built
1930	CG-2 glider	300
1941	P-7	1
1941	P-10	1
1942	CG-4A glider	750
1943	C-106A	2
1954-1962	CH-1 helicopter	40
1957	620	1
		1,095

Production History Cessna T-50 Bobcat

Model	1940	1941	1942	1943	1944	Total
T-50	19	21	3			43
AT-8		33				33
RCAF Crane I		588	52			640
RCAF Crane II			190			190
AT-17		24	985	131		1,140
UC-78			431	2459	466	3,356
Totals	19	666	1661	2590	466	5,402

Production History Cessna O-1 Birddog (L-19 Series)

Model	1950	1951	1952	1953	1954	1955	1956	1957	1959	1962-63	Total
O-1A (L-19A)	30	1186	671	318	195			99			2,499
OE-1 (Navy L-19)		30	17	13				2			62
L-19A-IT				66							66
O-1E (L-19E)							43	318	36	97	494
TO-1D (TL-19D)							100	207			307
XL-19B			1								1
XL-19C						2					2
Totals	30	1216	689	397	195	2	143	626	36	97	3,431

Cessna Model 150 "Commuter"

YEAR	MODEL	PRICE (F.A.F.)	ENGINE Continental	HP	GROSS WEIGHT	EMPTY WEIGHT	USEFUL LOAD	TOP SPEED	CRUISE SPEED	MAXIMUM RANGE (no res.)	RATE CLIMB
1959	150	$ 8,545	O-200A	100	1500 lbs	962 lbs	538 lbs	124 mph	121 mph	630 mi.	740
1960	150	8,795	O-200A	100	1500	962	538	124	121	630	740
1961	150A	8,995	O-200A	100	1500	950	550	124	121	630	740
1962	150B	8,995	O-200A	100	1500	950	550	127	125	610	760
1963	150C	9,435	O-200A	100	1500	950	550	127	125	610	760
1964	150D	9,495	O-200A	100	1600	1010	590	125	122	565	670
1965	150E	9,425	O-200A	100	1600	1010	590	125	122	565	670
1966	150F	9,275	O-200A	100	1600	1060	540	125	122	565	670
1967	150G	9,550	O-200A	100	1600	1060	540	125	122	565	670
1968	150H	10,195	O-200A	100	1600	1060	540	122	117	565	670
1969	150J	10,995	O-200A	100	1600	1060	540	122	117	565	670
1970	150K	11,450	O-200A	100	1600	1060	540	122	117	565	670
1970	A150K Aerobat	10,495	O-200A	100	1600	1020	580	120	115	565	670

Cessna Model 172 and Skyhawk

YEAR	MODEL	PRICE (F.A.F)	ENGINE	HP	GROSS WEIGHT	EMPTY WEIGHT	USEFUL LOAD	TOP SPEED	CRUISE SPEED	MAXIMUM RANGE (no res.)	RATE CLIMB
1956	172	$ 8,750	Continental O-300-A	145	2200 lbs	1260 lbs	940 lbs	135 mph	120 mph	600 mi.	660
1957	172	8,975	O-300-A	145	2200	1260	940	135	124	620	660
1958	172	8,995	O-300-A	145	2200	1260	940	135	124	620	660
1959	172	9,250	O-300-A	145	2200	1260	940	135	124	620	660
1960	172A	9,450	O-300-C	145	2200	1252	948	140	131	790	730
1961	Skyhawk (172B)	11,475	O-300-C & D	145	2200	1325	875	140	131	790	730
1962	Skyhawk (172C)	11,590	O-300-C & D	145	2250	1330	920	140	132	780	700
1963	Skyhawk (172D)	11,995	O-300-C & D	145	2300	1330	970	139	131	720	645
1963	Skyhawk Powermatic (P172D)	14,650	GO-300-E	175	2500	1425	1075	148	140	615	830
1964	Skyhawk (172E)	12,200	O-300-C & D	145	2300	1320	980	139	131	720	645
1965	Skyhawk (172F)	12,450	O-300-C & D	145	2300	1320	980	139	131	720	645
1966	Skyhawk (172G)	12,450	O-300-C & D	145	2300	1330	970	139	131	720	645
1967	Skyhawk (172H)	12,750	O-300-C & D	145	2300	1340	960	139	131	720	645
.968	Skyhawk (172I)	13,250	Lycoming O-320EZD	150	2300	1300	1000	140	132	655	645
.969	Skyhawk (172K)	13,995	O-320EZD	150	2300	1315	985	140	132	655	645
.970	Skyhawk (172L)	14,695	O-320EZD	150	2300	1315	985	140	132	655	645

Cessna 180 and Skywagon 180

YEAR	MODEL	PRICE (F.A.F.)	ENGINE Continental	HP	GROSS WEIGHT	EMPTY WEIGHT	USEFUL LOAD	TOP SPEED	CRUISE SPEED	MAXIMUM RANGE (no res.)	RATE CLIMB
1953	180	$12,950	0-470-A	225	2550 lbs	1480 lbs	1070 lbs	165+ mph	150+ mph	675 mi.	1150
1954	180	12,950	0-470-A	225	2550	1480	1070	165+	150+	675	1150
1955	180	12,950	0-470-J	225	2550	1480	1070	165+	150+	675	1150
1956	180	12,950	0-470-K	230	2550	1490	1060	165+	150+	675	1150
1957	180A	13,250	0-470-K	230	2650	1555	1095	170	160	675	1130
1958	180A	13,850	0-470-K	230	2650	1555	1095	170	160	845	1130
1959	180B	14,250	0-470-K	230	2650	1554	1096	170	160	845	1130
1960	180C	14,675	0-470-L	230	2650	1530	1120	170	160	845	1130
1961	180D	15,275	0-470-L	230	2650	1515	1135	170	160	845	1130
1962	180E	15,425	0-470-R	230	2650	1515	1135	170	162	925	1130
1963	180F	15,950	0-470-R	230	2650	1530	1120	170	162	925	1130
1964	180G	16,150	0-470-R	230	2800	1525	1275	170	162	925	1090
1965	180H	16,250	0-470-R	230	2800	1515	1285	170	162	925	1090
1966	180H	16,250	0-470-R	230	2800	1515	1285	170	162	925	1090
1967	180H	17,050	0-470-R	230	2800	1515	1285	170	162	925	1090
1968	180H	17,950	0-470-R	230	2800	1535	1265	170	162	925	1090
1969	Skywagon 180(180H)	18,850	0-470-R	230	2800	1545	1255	170	162	925	1090
1970	Skywagon 180(180H)	19,725	0-470-R	230	2800	1545	1255	170	162	925	1090

Cessna Model 185 and Skywagon 185

YEAR	MODEL	PRICE (F.A.F.)	ENGINE Continental	HP	GROSS WEIGHT	EMPTY WEIGHT	USEFUL LOAD	TOP SPEED	CRUISE SPEED	MAXIMUM RANGE (no res.)	RATE CLIMB
1961	Skywagon (185)	$18,950	IO-470-F	260	3200 lbs	1520 lbs	1680 lbs	176 mph	165 mph	885 mi	1000
1962	Skywagon (185A)	19,150	IO-470-F	260	3200	1520	1680	176	167	945	1000
1963	Skywagon (185B)	19,495	IO-470-F	260	3200	1580	1620	176	167	945	1000
1964	Skywagon (185C)	19,695	IO-470-F	260	3200	1560	1640	176	167	945	1000
1965	Skywagon (185D)	19,795	IO-470-F	260	3200	1560	1640	176	167	945	1000
1966	Skywagon (185E)	19,795	IO-470-F	260	3300	1560	1740	172	164	875	950
	Skywagon (A185E)	20,245	IO-520D	300	3350	1570	1730	178	169	825	1040
1967	Skywagon (A185E)	20,395	IO-520D	300	3350	1565	1735	178	169	825	1040
1968	Skywagon (A185E)	21,375	IO-520D	300	3350	1575	1775	178	169	825	1010
1969	Skywagon 185 (A185E)	22,450	IO-520D	300	3350	1575	1775	178	169	825	1010
1970	Skywagon 185 (A185E)	23,350	IO-520D	300	3350	1575	1775	178	169	825	1010

Cessna Model 182 and Skylane

YEAR	MODEL	PRICE (F.A.F.)	ENGINE Continental	HP	GROSS WEIGHT	EMPTY WEIGHT	USEFUL LOAD	TOP SPEED	CRUISE SPEED	MAXIMUM RANGE (no res.)	RATE CLIMB
1956	182	$13,750	0-470-L	230	2550 lbs	1510 lbs	1040 lbs	160+ mph	150+ mph	675+ mi.	1100
1957	182A	13,975	0-470-L	230	2650	1560	1090	165	155	805	1030
1958	Skylane (182B)	16,850	0-470-L	230	2650	1560	1090	168	158	819	1030
1959	Skylane (182B)	17,095	0-470-L	230	2650	1594	1056	170	160	845	1030
1960	Skylane (182C)	17,325	0-470-L	230	2650	1632	1018	170	160	845	1030
1961	Skylane (182D)	17,950	0-470-L	230	2650	1615	1035	170	160	845	1030
1962	Skylane (182E)	18,490	0-470-R	230	2800	1625	1175	170	162	925	980
1963	Skylane (182F)	18,990	0-470-R	230	2800	1625	1175	170	162	925	980
1964	Skylane (182G)	17,875	0-470-R	230	2800	1610	1190	170	162	925	980
1965	Skylane (182H)	17,995	0-470-R	230	2800	1610	1190	170	162	925	980
1966	Skylane (182J)	17,995	0-470-R	230	2800	1620	1180	170	162	925	980
1967	Skylane (182K)	18,450	0-470-R	230	2800	1620	1180	170	162	925	980
1968	Skylane (182L)	18,995	0-470-R	230	2800	1625	1175	170	162	925	980
1969	Skylane (182M)	19,950	0-470-R	230	2800	1635	1165	170	162	925	980
1970	Skylane (182N)	20,895	0-470-R	230	2950	1640	1310	168	160	885	890

Cessna Model T/P-206 and Super Skylane

YEAR	MODEL	PRICE (F.A.F.)	ENGINE Continental	HP	GROSS WEIGHT	EMPTY WEIGHT	USEFUL LOAD	TOP SPEED	CRUISE SPEED	MAXIMUM RANGE (no res.)	RATE CLIMB
1963	205	$22,295	IO-470-S	260	3300 lbs	1750 lbs	1550 lbs	173 mph	163 mph	1015 mi.	965
1964	205A	22,525	IO-470-S	260	3300	1750	1550	173	163	1015	965
1965	Super Skylane (P206)	23,995	IO-520-A	285	3300	1790	1510	177	166	825	1075
1966	Super Skylane (P206A)	23,995	IO-520-A	285	3600	1785	1815	174	163	810	920
	T/S Super Skylane (TP206A)	26,995	TSIO-520-C	285	3600	1885	1715	200	184	800	1030
1967	Super Skylane (P206B)	23,995	IO-520-A	285	3600	1795	1805	174	163	810	920
	T/S Super Skylane (TP206B)	27,995	TSIO-520-C	285	3600	1895	1705	200	184	825	1030
1968	Super Skylane (P206C)	24,995	IO-520-A	285	3600	1820	1780	174	163	810	920
	T/S Super Skylane (TP206C)	29,195	TSIO-520-C	285	3600	1915	1685	200	184	825	1030
1969	Super Skylane (P206D)	26,550	IO-520-A	285	3600	1835	1765	174	163	810	920
	T/S Super Skylane (TP206D)	30,950	TSIO-520-C	285	3600	1935	1665	200	184	825	1030
1970	Super Skylane (P206E)	27,450	IO-520-A	285	3600	1835	1765	174	163	810	920
	T/S Super Skylane (TP206E)	31,950	TSIO-520-C	285	3600	1935	1665	200	184	825	1030

Cessna Model T/U-206 and Skywagon 206

YEAR	MODEL	PRICE (F.A.F.)	ENGINE Continental	HP	GROSS WEIGHT	EMPTY WEIGHT	USEFUL LOAD	TOP SPEED	CRUISE SPEED	MAXIMUM RANGE (no res.)	RAT CL
1964	Super Skywagon (20[illegible])	$22,950	IO-520-A	285	3300 lbs	1780 lbs	1520 lbs	177 mph	166 mph	825 mi.	107
1965	Super Skywagon (U206)	22,950	IO-520-A	285	3300	1760	1540	177	166	825	107
1966	Super Skywagon (U206A)	22,950	IO-520-A	285	3600	1690	1910	174	163	810	920
1966	Turbo S/SW (TU-206A)	25,950	TSIO-520-C	285	3600	1785	1815	200	184	825	103
1967	Super Skywagon (U-206B)	22,950	IO-520-F	285*	3600	1695	1905	174	164	800	920
1967	Turbo S/SW (TU-206B)	26,950	TSIO-520-C	285	3600	1795	1805	200	184	825	103
1968	Super Skywagon (U-206C)	23,850	IO-520-F	285*	3600	1710	1890	174	164	800	920
1968	Turbo S/SW (TU-206C)	28,050	TSIO-520-C	285	3600	1795	1805	200	184	825	103
1969	Super Skywagon (U-206D)	25,275	IO-520-F	285*	3600	1725	1875	174	164	800	920
1969	Turbo S/SW (TU-206D)	29,675	TSIO-520-C	285	3600	1825	1775	200	184	825	103
1970	Skywagon 206 (U-206E)	25,995	IO-520-F	285*	3600	1710	1890	174	164	800	920
1970	Turbo System Skywagon 206 (TU-206E)	30,495	TSIO-520-C	285	3600	1810	1790	200	184	825	103

* 300 HP at takeoff; 285 HP continuous

Cessna 177 and Cardinal

YEAR	MODEL	PRICE (F.A.F.)	ENGINE Lycoming	HP	GROSS WEIGHT	EMPTY WEIGHT	USEFUL LOAD	TOP SPEED	CRUISE SPEED	MAXIMUM RANGE (no res.)	RATE CLIMB
1968	Cardinal (177)	$14,500	0-320-EZD	150	2350 lbs	1415 lbs	935 lbs	144 mph	134 mph	855 mi.	670
1969	Cardinal (177)	16,995	0-360-AZF	180	2500	1440	1060	150	138	770	760
1970	Cardinal (177A)	17,500	0-360-AZF	180	2500	1475	1025	153	142	790	840

Cessna Model 207 and Skywagon 207

YEAR	MODEL	PRICE (F.A.F.)	ENGINE Continental	HP	GROSS WEIGHT	EMPTY WEIGHT	USEFUL LOAD	TOP SPEED	CRUISE SPEED	MAXIMUM RANGE (no res.)	RATE CLIMB
1969	Skywagon 207	$28,250	IO-520-F	285*	3800 lbs	1860 lbs	1940 lbs	168 mph	159 mph	695 mi.	810
	Turbo Skywagon T207	32,975	TSIO-520-G	285*	3800	1960	1840	189	176	685	885
1970	Skywagon 207	28,250	IO-520-F	285*	3800	1880	1920	168	159	695	810
	Turbo Skywagon T207	32,975	TSIO-520-G	285*	3800	1980	1820	189	176	685	885

*Rated at 300 HP at takeoff; 285 HP continuous

Cessna Model 210 Centurion and T210 Centurion

YEAR MODEL	PRICE (F.A.F.)	ENGINE Continental	HP	GROSS WEIGHT	EMPTY WEIGHT	USEFUL LOAD	TOP SPEED	CRUISE SPEED	MAXIMUM RANGE (No res.)	RATE CL
1960 210	$22,450	IO-470E	260	2900 lbs	1735 lbs	1165 lbs	199 mph	190 mph	1100 mi.	130
1961 210A	23,450	IO-470E	260	2900	1740	1160	199	190	1075	130
1962 210B	23,975	IO-470-S	260	3000	1750	1250	198	189	1215	127
1963 210C	24,625	IO-470-S	260	3000	1780	1220	198	189	1215	127
1964 Centurion (210D)	25,750	IO-520-A	285	3100	1830	1270	199	191	1035	121
1965 Centurion (210E)	25,975	IO-520-A	285	3100	1860	1240	199	191	1035	121
1966 Centurion (210F)	25,975	IO-520-A	285	3300	1865	1435	198	190	1010	111
Turbo System Centurion (T210F)	28,975	TSIO-520-C	285	3300	1965	1335	230	220	1110	128
1967 Centurion (210G)	27,975	IO-520-A	285	3400	1960	1440	200	192	980	100
Turbo System Centurion (T210G)	31,975	TSIO-520-C	285	3400	2050	1350	234	223	1010	11
1968 Centurion (210H)	29,475	IO-520-A	285	3400	1960	1440	200	192	980	10
Turbo System Centurion (T210H)	33,675	TSIO-520-C	285	3400	2050	1350	234	223	1010	11
1969 Centurion (210J)	31,250	IO-520-J	285	3400	1960	1440	200	192	980	10
Turbo System Centurion (T210J)	35,695	TSIO-520-H	285	3400	2060	1340	234	223	1010	11
1970 Centurion (210K)	34,725	IO-520-L	285	3800	2080	1720	200	188	1250	86
Turbo System Centurion (T210K).	39,250	TSIO-520-H	285	3800	2180	1620	230	219	1330	93

Cessna Agwagon "B"

THE CESSNA AGWAGON "B"

The Agwagon "B," designed for dispersal of liquid or dry chemicals, can carry over 1,900 lbs of useful load, has a hopper capacity of 200 gallons and works at swath speeds from 85 to 120 mph.

The Agwagon was introduced in 1966, and improvements accruing through the 1970 Model included a re-designed hopper to allow quicker loading and dumping, a variety of dispersal equipment options and all-around visibility for the pilot, plus a pair of tinted top windows to permit additional pilot-vision while turning at the end of dispersal runs. The "B" Model also had a more rugged, thicker main-gear, increased dihedral, wing fences and aileron leading-edge gap seals.

Dispersal systems accessories included an engine-driven hydraulic spray system, an industry first. Exclusive on the Agwagon, the internally-mounted pump conserves engine horsepower and reduces drag for improved aircraft performance. Other dispersal options were an electrically-operated spray valve for remote pump spray systems, and a spring-loaded check valve to prevent dry material in the hopper from entering the optional sideloading plumbing.

Standard equipment on the Agwagon "B" included a universal dispersal control system, bullet spinner for constant-speed prop installation, control stick lock, navigation lights, retractable tailcone lift handles and quick-release canopy door hinges.

Servicing was simplified with a quickly removable engine air filter and a remote quick-drain for the oil.

Factory list price of the 1970 Agwagon "B" was $19,400.

Cessna AGWAGON "B"

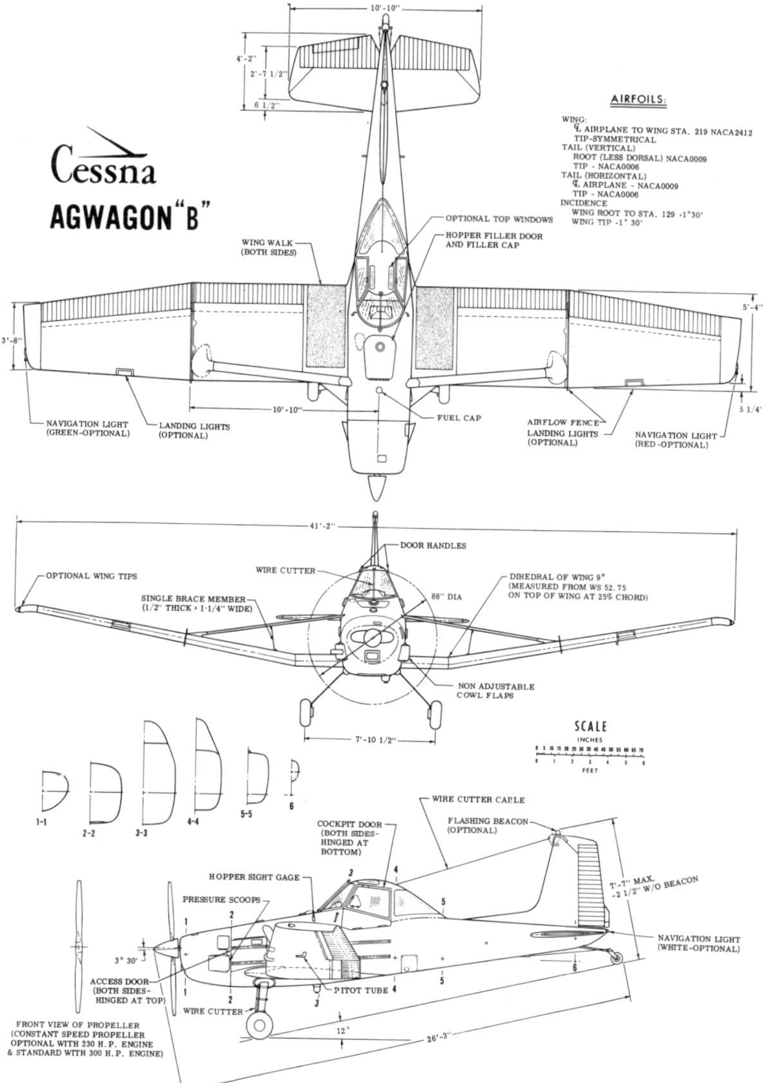

CESSNA AIRCRAFT CO. WICHITA, KANSAS

1970 Performance and Specifications—Agwagon "B"

Liquid Dispersal System With Engine-Driven Hydraulic Pump

	230 hp, Fixed Pitch	230 hp, Constant Speed	300 hp, Constant Speed
GROSS WEIGHT:	3300 lbs,, 1497 kg	3300 lbs, 1497 kg	3300 lbs, 1497 kg
SPEED: Best Power Mixture Top Speed at Sea Level	113 mph, 182 kph at rated 2600 rpm	125 mph, 201 kph at rated power	134 mph, 216 kph at max continuous powe
Cruise Speed	75% power at 5000' 107 mph, 172 kph	75% power at 6500' 114 mph, 183 kph	75% power at 6500' 127 mph, 204 kph
RANGE: Normal Lean Mixture 36.5 Gallons, No Reserve	75% power at 5000' 280 miles, 451 km 2.6 hrs. 107 mph, 172 kph	75% power at 6500' 295 miles, 475 km 2.6 hrs. 114 mph, 183 kph	75% power at 6500' 290 miles, 467 km 2.3 hrs. 126 mph, 202 kph
54 Gallons, No Reserve			430 miles, 692 km 3.4 hrs. 126 mph, 202 kph
RATE OF CLIMB AT SEA LEVEL:	590 fpm, 180 mpm	635 fpm, 194 mpm	820 fpm, 250 mpm
SERVICE CEILING:	10,600 ft, 3231 m	11,100 ft, 3383 m	13,500 ft, 4115 m
TAKEOFF: Ground Run	960 ft, 293 m	860 ft, 262 m	630 ft, 192 m
Total Distance Over 50' Obstacle	1545 ft, 471 m	1420 ft, 433 m	1000 ft, 305 m
LANDING: Ground Roll	420 ft, 128 m	420 ft, 128 m	420 ft, 128 m
Total Distance Over 50' Obstacle	1265 ft, 386 m	1265 ft, 386 m	1265 ft, 386 m
STALL SPEED: Flaps Up, Power Off	62 mph 100 kph	62 mph 100 kph	62 mph 100 kph
Flaps Down, Power Off	58 mph, 93 kph	58 mph, 93 kph	58 mph, 93 kph
EMPTY WEIGHT: (Approximate)	1910 lbs, 866 kg	1925 lbs, 873 kg	1940 lbs, 880 kg
MAXIMUM GROSS WEIGHT: (Restricted Category)	3800 lbs, 1722 kg	3800 lbs, 1722 kg	4000 lbs, 1814 kg
GROSS WEIGHT:	3300 lbs, 1497 kg	3300 lbs, 1497 kg	3300 lbs, 1497 kg
SPEED: Best Power Mixture Top Speed at Sea Level	108 mph, 172 kph at rated 2600 rpm	112 mph, 180 kph at rated power	121 mph, 195 kph at max continuous pow
Cruise Speed	75% power at 5000' 95 mph, 153 kph	75% power at 6500' 101 mph, 163 kph	75% power at 6500' 114 mph, 183 kph
RANGE: Normal Lean Mixture 36.5 Gallons, No Reserve	75% power at 5000' 245 miles, 394 km 2.6 hrs. 95 mph, 153 kph	75% power at 6500' 260 miles, 418 km 2.6 hrs. 101 mph, 163 kph	75% power at 6500' 260 miles, 418 km 2.3 hrs. 113 mph, 182 kph
54 Gallons, No Reserve			390 miles, 628 km 3.4 hrs. 113 mph, 182 kph
RATE OF CLIMB AT SEA LEVEL:	460 fpm, 140 mpm	505 fpm, 154 mpm	690 fpm, 210 mpm
SERVICE CEILING:	7700 ft, 2347 m	8400 ft, 2560 m	11,100 ft, 3383 m
TAKEOFF: Ground Run	1040 ft, 317 m	940 ft, 287 m	680 ft, 207 m
Total Distance Over 50' Obstacle	1740 ft, 530 m	1620 ft, 494 m	1090 ft, 332 m
LANDING: Ground Roll	420 ft, 128 m	420 ft, 128 m	420 ft, 128 m
Total Distance Over 50' Obstacle	1265 ft, 386 m	1265 ft, 386 m	1265 ft, 386 m
STALL SPEED: Flaps Up, Power Off	62 mph, 100 kph	62 mph, 100 kph	62 mph, 100 kph
Flaps Down, Power Off	58 mph, 93 kph	58 mph, 93 kph	58 mph, 93 kph
EMPTY WEIGHT: (Approximate)	1940 lbs, 880 kg	1955 lbs, 887 kg	1970 lbs, 894 kg
MAXIMUM GROSS WEIGHT: (Restricted Category)	3800 lbs, 1722 kg	3800 lbs, 1722 kg	4000 lbs, 1814 kg